Study Guide

for

Stark's

SOCIOLOGY

Internet Edition

Ninth Edition

Study Guide

for

Stark's

SOCIOLOGY

Internet Edition

Ninth Edition

Robert A. Wortham
North Carolina Central University

Australia • Canada • Mexico • Singapore • Spain • United Kingdom • United States

Printed in the United States of America
1 2 3 4 5 6 7 07 06 05 04 03

ISBN: 0-534-60942-2

Printer: Globus Printing

For more information about our products, contact us at:
Thomson Learning Academic Resource Center
1-800-423-0563

For permission to use material from this text, contact us by:
Phone: 1-800-730-2214
Fax: 1-800-731-2215
Web: http://www.thomsonrights.com

Cover Image: "Painters on Suspenders, October 7, 1914," by Bridge Department photographer Eugene de Salignac. Municipal Archives Photo Collection, New York City Department of Records and Services. Reprinted by permission.

Wadsworth-Thomson Learning, Inc.
10 Davis Drive
Belmont, CA 94002-3098
USA

Asia
Thomson Learning
5 Shenton Way #01-01
UIC Building
Singapore 068808

Australia/ New Zealand
Thomson Learning
102 Dodds Street
Southbank, Victoria 3006
Australia

Canada
Nelson
1120 Birchmount Road
Toronto, Ontario M1K 5G4
Canada

Europe/Middle East/South Africa
Thomson Learning
High Holborn House
50/51 Bedford Row
London WC1R 4LR
United Kingdom

Latin America
Thomson Learning
Seneca, 53
Colonia Polanco
11560 Mexico D.F.
Mexico

Spain/ Portugal
Paraninfo
Calle/Magallanes, 25
28015 Madrid, Spain

CONTENTS

PREFACE

The Stark Approach

Rodney Stark's approach to the study of sociology is distinctive. As you read the ninth edition of *Sociology*, you will not encounter a general "review of the literature" approach to basic sociological topics like socialization, stratification, culture and deviance. Rather, you will be given an opportunity to engage in what C. Wright Mills termed "the sociological imagination." You will learn about what sociology "is" by observing what sociologists "do." You will be exposed to major sociological theories, and you will gain an appreciation for the empirical method. This approach will enable you to see why data are an integral part of the sociological process. In each chapter a few major concepts are introduced, developed and tested so that you can see what sociology is and how sociologists use these concepts to gain a better understanding of everyday life experiences. This can be a refreshing, exciting and challenging journey. Welcome aboard!

The Study Guide: Purpose and Function

This *Study Guide* is designed to help you gain a better understanding of Stark's distinctive approach to the sociological process. Since Stark's approach is highly conceptual and empirical, you may find that you will have to "dig a little deeper" with the first two or three chapters before you begin to feel comfortable with this approach. To help you through this transition, an extended abstract and a detailed chapter outline accompanies each chapter. Please treat them as conceptual road maps. Each chapter will also include a listing of key terms and brief summaries of the major empirical studies presented. So that you may stay current with some of the recent developments in the field, *Info Trac* search terms are included with each chapter review. These search terms will enable you to "do" a little sociology on your own. As with any study guide, review questions are provided for each chapter. Answers and page references are provided in most instances. Many of the questions will test your general mastery of the subject matter, but others will encourage you to enhance your critical-thinking skills.

The *Study Guide* contains twenty-one chapters and follows the main text's chapter sequence. Features included with each chapter review are:

- An Extended Abstract
- 15 Key Learning Objectives
- A Detailed Chapter Outline
- A List of Key Terms with Page References
- Key Research Study Summaries with Page References
- 5 *Info Trac* Search Words (2 Identified for More Extensive Investigation)
- 30 Multiple Choice Questions
- 10 True/False Questions
- 10 Short Answer Questions
- 5 Essay Questions
- Answer Sheet

Please remember that the *Study Guide* is primarily a tool that will enable you to gain a better understanding of the material included in the Stark text. It is not intended as a substitute for reading the text!

Web Quizzes

Please consult the main text's Web site (www.socstark9.com) if you would like to take advantage of another opportunity to test your understanding of the material presented in each chapter. Here, I have provided twenty additional multiple choice questions and an answer key for each chapter.

Acknowledgements

Working on this *Study Guide* has been exciting and challenging. As with any project of this magnitude, the help and assistance received from others is deeply appreciated. The contributions of Carol Barbee Wortham, Assistant Professor of Sociology at Shaw University, are extensive. Professor Carol Wortham has graciously served as the copy editor for the *Study Guide* and Web Quizzes. She also assisted with much of the initial typing. Her dedication to each project has been a source of inspiration, and her sociological insights have enabled me to produce manuscripts that are more readable. Thanks are extended also to Stephanie Monzon and Natalie Cornelison of Wadsworth Publishing Company. The flexibility extended in meeting important deadlines has been most helpful and deeply appreciated. Finally, I would like to thank Coy Horton and the late Erline Horton for reminding me that strong social networks are the fabric of daily life. People are far more important than things.

Robert A. Wortham

CHAPTER 1

GROUPS AND RELATIONSHIPS: A SOCIOLOGICAL SAMPLER

Extended Abstract

Stark lays the foundation for the sociological imagination in this introductory chapter. Since sociologists attempt to provide testable explanations of observable everyday life phenomena, theory and research represent two critical cornerstones of the sociological imagination. Early nineteenth century moral statisticians such as Guerry and Morselli offered empirical descriptions of social phenomena as they investigated the stability, geographic variation and increase in nineteenth century European suicide rates. Durkheim offered a more formal explanation of these trends as he linked suicide rates to the strength of a society's social networks.

Next, the ties among sociology and the social sciences are noted. Stark proceeds to develop key methodological concepts such as units of analysis and the microsociology / macrosociology distinction. The unit of analysis can be individuals, groups or ecological units like states or nations. Microsociological studies address individual and small group interactions; whereas, macrosociological studies address larger units of analysis such as cross-national comparisons.

Distinctions between groups and aggregates and primary and secondary groups are offered next. Groups provide a sense of identity and are maintained through stable relationships, which may be of a more intimate (primary group) or formal (secondary group) nature. Social interactions may encourage a sense of common identity or social solidarity as well as promote conflict. Sampson's study on monastic life is offered as a case study in social network analysis.

The remainder of this sociological primer includes discussions of data collection techniques, research checks and balances, the key steps of the sociological process and the question of free will. Stark notes that sometimes people are aware that they are being studied. Since self-aware subjects can produce biased responses, unobtrusive measures based on behavioral traces are often constructed. Stark provides a "look over his shoulder" as he demonstrates how he and a colleague constructed an unobtrusive measure for astrology using data from the *Yellow Pages*. He then demonstrates how magazine subscription data, another unobtrusive measure, were used to validate UFO data obtained from newspaper reports.

Two important assumptions concerning empirical research are introduced next. First, it is assumed that researchers will separate personal beliefs (bias) and research findings. Second, within a given social environment, persons can exercise free will. The sociologist's task is to determine how and under what conditions choices are exercised.

In concluding this sociological primer, Stark notes that the sociological process includes theory construction and theory testing. This framework is adopted throughout the book as a means of presenting and evaluating key sociological concepts. Finally, within this introductory chapter, a brief overview of the growth and development of the sociological tradition in the U.S. is provided. The contributions of Albion Small (the Chicago tradition) and W.E.B. Du Bois (the Atlanta tradition) are noted.

Key Learning Objectives

After a thorough reading of Chapter 1, you should be able to:

1. distinguish theory and research as two major cornerstones of empirical science.
2. understand how the research of the moral statisticians and Durkheim illustrates the discovery of social facts and the stimulation of the sociological imagination.
3. explain why sociology is grouped with the social sciences.
4. look at statistical data presented in table format and draw appropriate conclusions.
5. identify a study's unit of analysis.
6. understand how microsociological and macrosociological approaches focus on different units of analysis.
7. gain an appreciation for the variety of sources of sociological data available for investigation.
8. distinguish primary and secondary groups and know how social network analysis can be a valuable tool for identifying social solidarity and social conflict within groups.
9. realize how self-aware measures can be problematic.
10. know how behavioral data are obtained using unobtrusive measures and how different data sources may be used to validate research findings.
11. appreciate the need to control for researcher bias and identify the various sources of researcher bias.
12. identify the major steps in the sociological process.
13. understand the association between theory construction and theory testing.
14. identify the complex association between free will and a fatalistic approach to human behavior.
15. appreciate the distinct contributions of Small and Du Bois to the development of the U. S. sociological tradition.

Chapter Outline

I. The Cornerstones of Sociological Inquiry
   A. Theory and Research
      1. Theories provide abstract but testable explanations of observable phenomena.
      2. Research involves data collection and data analysis.
   B. Moral Statistics and the Sociological Imagination.
      1. In 1827 the French Ministry of Justice published a comprehensive data source, *Compte*, on arrests and convictions.
      2. Guerry and Quetelet observe stability and variation in suicide rates.
      3. Morselli ties variation and increase in suicide rates to modernization.
      4. Durkheim links variation in suicide rates to strength of social relationships.
      5. The sociological imagination involves understanding how social forces shape individual actions.
   C. Sociology and the Social Sciences
      1. The Social sciences provide a general understanding of the various patterns of human behavior.
      2. Sociology investigates the social influences on human behavior.
      3. Sociology is an integrated science of society.

II. Key Sociological Concepts
   A. Units of Analysis
      1. May be individuals, groups, states or nations.
      2. Abortion data used to illustrate different units of analysis.
   B. Microsociology and Macrosociology
      1. Microsociology addresses interactions among individuals and small groups.
      2. Macrosociology addresses comparisons among larger groups such as states.
   C. Groups versus Aggregates
      1. Groups maintain stable patterns of relationships.
      2. Aggregates involve accidental, brief encounters among persons.
      3. Primary group ties are more intimate and emotional.
      4. Secondary group ties are more formal and task oriented.
   D. Social Network Analysis
      1. Strong social ties enhance social solidarity.
      2. Social conflict may challenge social solidarity.
      3. Social network analysis may be used to study social solidarity and social conflict.
      4. Sampson's study on monastic strife is grounded in social network analysis.
   E. Self-Aware versus Unobtrusive Measures
      1. Self-aware responses may be biased.

2. Unobtrusive measures are based on behavioral traces.
3. Stark and Bader use *Yellow Page* data and newspaper clippings to construct unobtrusive measures.

F. Validation Research
1. Use of other data sets as reference standards.
2. Use of similar measures as reference standards.
3. Stark and Bader use magazine subscription rates to validate newspaper data.

G. Researcher Bias
1. Separate personal beliefs from research findings.
2. Research findings are tested and retested.
3. Researchers should provide full disclosure of methodology and findings.

III. The Sociological Process

A. Theory Construction and Theory Testing
1. Theory construction (wonder, conceptualize and theorize).
2. Theory testing (operationalize, hypothesize, observe, analyze and assess).

B. Empirical Research Assumptions
1. Humans can make rational choices and possess free will.
2. Social sciences identify conditions under which choices are exercised.

IV. Development of Sociology in the U.S.

A. Albion Small and the Chicago Tradition
1. Founded the first U.S. sociology department.
2. Wrote a sociology textbook and founded a major sociological journal.

B. W.E.B. Du Bois and the Atlanta Tradition
1. Established sociological laboratory at Atlanta University.
2. Published numerous sociological studies on African American experiences in the U.S.
3. Strong advocate of research objectivity.

Key Terms

Based on your reading of Chapter 1, you should be able to define and illustrate the key sociological concepts listed below. Page numbers are provided in parentheses as reference points.

Science (2)
Theory (2)
Empirical (2)
Research (2)
Moral Statistics (4)
Sociological Imagination (6)
Social Facts (7)
Sociology (7)
Social Sciences (7)
Unit of Analysis (9)
Micro Sociology (10)
Macro Sociology (12)
Global Perspective (12)
Concept (13)
Group (13)
Aggregate (13)
Primary Group (13)

Secondary Group (14)
Social Solidarity (14)
Social Conflict (14)
Network (15)
Social Network (15)
Social Relationship (15)
Unobtrusive Measures (18)
Validation Research (19)
Self-Report Data (19)
Systematic Skepticism (20)
Bias (20)
The Sociological Process (21)
Conceptualize (22)
Operationalize (22)
Hypothesis (22)

Religious Determinism (22)
Free Will (22)

Key Research Studies

Listed below are key research studies cited in Chapter 1. Familiarize yourself with the major findings of these studies. Page references are provided in parentheses.

Homans' law of inequality: friendship ties generally created and maintained by persons of the same rank (2).

Moral statisticians (Guerry, Quetelet, Wagner and Morselli): stability, variation and increase in nineteenth century European suicide rates (4-6).

Durkheim: *Suicide* (1897) – classic sociological study linking suicide rates and strength of social ties (6-7).

Sampson: use of social network analysis to study social conflict among monastery cliques (16).

Stark and Bader: construction of unobtrusive measures to measure interest in astrology and belief in UFOs (17-20).

Info Trac Search Words

Enter these search terms to conduct more extensive investigations of key topics introduced in Chapter 1.

Empirical Method

Macrosociology

Suicide Studies

Look at one of the articles dealing with suicide and note how the data for the study were collected and analyzed. Identify the important factors associated with variations in suicide rates. How do the findings of your chosen study compare with Durkheim's?

General Social Survey

Stark uses this data source throughout the text. Select one of the articles cited and identify the study's unit of analysis. Is the study primarily a microsociological or macrosociological study?

Free Will and Determinism
Subdivisions: Social Aspects

Multiple Choice

Answers and page references are provided at the chapter end.

1. The two major cornerstones of the scientific method are
   a. fatalism and free will.
   b. theory and research.
   c. choice and bias.
   d. theory and free will.
   e. bias and research.

2. A theory
   a. must have empirical implications.
   b. is an abstract statement explaining how events take place.
   c. is an abstract statement explaining how things really are.
   d. all of the above
   e. none of the above

3. According to Homan's law of inequality, friendships tend to be concentrated
   a. more among women than men.
   b. more among younger than older people.
   c. among people of the same rank.
   d. among persons of different rank.
   e. among primary rather than secondary group members.

4. In *The Wealth of Nations*, Smith developed his
   a. economic theory of supply and demand.
   b. anthropological theory of cultural lag.
   c. psychological theory about the nature of personality.
   d. sociological theory of conflict.
   e. political theory about the power elite.

5. The *Compte* is an early example of
   a. a sociological theory.
   b. moral statistics.
   c. researcher bias.
   d. an hypothesis.
   e. individual facts.

6. According to Morselli the upward trend in nineteenth century European suicide rates could be attributed to
   a. small town environments.
   b. the growth of rural societies.
   c. an increased reliance on agricultural employment.
   d. increased industrialization and modernization.
   e. the widespread decline in crime rates.

7. In his classic sociological study on suicide, Durkheim observed that high suicide rates were associated with
   a. agricultural employment.
   b. rural rather than urban environments.
   c. weak social networks.
   d. the Protestant Reformation.
   e. low crime rates.

8. Each of the following areas of study is known as a social science except
   a. psychology.
   b. anthropology.
   c. music.
   d. political science.
   e. sociology.

9. An appropriate unit of analysis can be a/an
   a. individual.
   b. group.
   c. state.
   d. nation.
   e. all of the above

10. This sociologist established the first sociology department in the U.S. and founded the *American Journal of Sociology*.
   a. Albion Small
   b. Emile Durkheim
   c. Rodney Stark
   d. W.E.B. Du Bois
   e. Adam Smith

11. This early U.S. sociologist established a sociological laboratory at Atlanta University and investigated social forces influencing African American life.
   a. Andre Michel Guerry
   b. Henry Morselli
   c. Karl Marx
   d. W.E.B. Du Bois
   e. August Comte

12. The *World Values Survey* is an example of
  a. social network analysis.
  b. an empirical data set.
  c. free will.
  d. bias.
  e. a theory.

13. Microsociology is to macrosociology as
  a. first is to second.
  b. free is to imprisoned.
  c. self-aware data is to unobtrusive data.
  d. primary is to secondary.
  e. small is to large.

14. States are often used as the unit of analysis in
  a. macrosociological studies.
  b. microsociological studies.
  c. primary sociological studies.
  d. secondary sociological studies.
  e. none of the above

15. Sociologists argue that a group must include at least
  a. one person.
  b. two persons.
  c. three persons.
  d. four persons.
  e. five persons.

16. An accidental, brief collection of people best represents a/an
  a. primary group.
  b. secondary group.
  c. aggregate.
  d. clique.
  e. reference group.

17. Morselli and Durkheim believed that modernization would
  a. strengthen primary group ties.
  b. weaken primary group ties.
  c. not affect primary group ties.
  d. would strengthen aggregates.
  e. would weaken aggregates.

18. The family is generally treated as a/an
   a. secondary group.
   b. unobtrusive group.
   c. primary group.
   d. self-aware group.
   e. validation group.

19. Business organizations and social clubs are often identified as examples of
   a. cliques.
   b. aggregates.
   c. secondary groups.
   d. loyalist groups.
   e. primary groups.

20. As the sense of belonging and loyalty increases, social solidarity tends to
   a. increase.
   b. decrease.
   c. remain the same.
   d. decrease initially and then increase.
   e. increase initially and then decrease.

21. Sociologists treat unfriendly interaction between groups as examples of
   a. aggregates.
   b. unobtrustive measures.
   c. social integration.
   d. social solidarity.
   e. social conflict.

22. Social network analysis may be used to
   a. measure social relationships.
   b. diagram social relationships.
   c. explore social solidarity.
   d. investigate social conflict.
   e. all of the above

23. In the study on monastic life, the older men who were satisfied with the existing monastic rules and wanted to keep things the same were labeled
   a. outcasts.
   b. loyalists.
   c. rebels.
   d. monks.
   e. leaders.

24. If a researcher used a questionnaire to study alcohol use among college students, the researcher would be
    a. studying subjects who are self-aware.
    b. constructing an unobtrusive measure.
    c. developing a primary measure.
    d. developing a secondary measure.
    e. stating a hypothesis.

25. Instead of using a questionnaire to measure belief in astrology, Stark and Bader decided to use astrology listings from the *Yellow Pages*. This is an example of
    a. theory construction.
    b. moral statistics.
    c. an unobtrusive measure.
    d. researcher bias.
    e. self-aware data.

26. A major concern of validation research is
    a. social network analysis.
    b. verifying data accuracy.
    c. producing unobtrusive measures.
    d. collecting moral statistics.
    e. minimizing fatalism.

27. Self-report crime data may be checked against court records as a way of improving the accuracy of the self-report data. This is an example of
    a. testing hypotheses.
    b. theory construction.
    c. validation research.
    d. use of an unobtrusive measure to verify self-report data.
    e. c and d

28. A researcher's personal opinions can influence a study's outcome. Personal opinions are an example of
    a. social conflict.
    b. social networks.
    c. researcher bias.
    d. moral statistics.
    e. modernization.

29. Theory testing includes each of the following steps in the sociological process except
    a. operationalize.
    b. wonder.
    c. hypothesize.
    d. observe.
    e. analyze.

30. According to the doctrine of free will
  a. all human actions are preordained.
  b. human beings cannot alter their fates.
  c. humans are able to make reasonable choices.
  d. human actions are determined by the gods.
  e. all human choices are irrational.

True/False

Answers and page references are provided at the chapter end.

1. The goal of research is to test theories and gain sufficient knowledge.

2. Morselli noted that nineteenth century European suicide rates reflected a continuous declining pattern.

3. Sociologists are primarily concerned with studying the actions of groups.

4. A sociologist who uses the state as the unit of analysis is probably conducting a macrosociological study.

5. A group may be defined as a brief, accidental gathering of individuals.

6. Primary groups are more intimate groups and tend to play a major role in shaping identity.

7. In social network analysis, diagrams are often used to identify the patterns of social relationships that exist among group members.

8. Unobtrusive measures are used when a researcher wants to gain information from persons who are aware that they are being studied.

9. If a researcher wanted to check data on the strength of African American social networks obtained from the *General Social Survey* with data on social networks obtained from the *National Survey of Black Americans*, the researcher would be attempting to validate the social network pattern noted in the *General Social Survey* data.

10. In the sociological process model, theory testing follows theory construction.

Short Answer Questions

These short answer questions are provided to test your knowledge and understanding of the basic sociological concepts presented in Chapter 1. Page references for answers are included at the chapter end.

1. According to Homans' law of inequality, what is the nature of the association between friendship ties and social rank?

2. Why was the *Compte* an example of an important early source of empirical data?

3. How do microsociological studies differ from macrosociological studies with regard to identification of the unit of analysis?

4. Identify three reasons why modern sociologists may choose to frame their research within a global perspective.

5. How might primary and secondary groups be distinguished?

6. Why might a sociologist employ social network analysis to study small group behavior?

7. What is a major problem researchers face when their subjects are aware that they are being studied?

8. Why do researchers conduct validation research?

9. What is the meaning of the following statement, "The essence of the scientific method is systematic skepticism?"

10. What is a research hypothesis?

Essay Questions

These questions are designed to test your understanding of key sociological concepts presented in Chapter 1 and your ability to apply these insights to concrete situations.

1. How do the studies of suicide by the moral statisticians and Durkheim illustrate how the scientific method may be applied to the study of a social phenomenon?

2. Provide several examples of how secondary groups can function as primary groups in times of crisis.

3. Explain how someone could employ social network analysis to gain a better understanding of the interpersonal dynamics involved in sorority and fraternity pledge periods.

4. How does researcher bias involve personal bias and methodological bias?

5. Design a research project that will investigate drug use among college students using the eight steps of the sociological process as your methodological framework.

Answers

Multiple Choice

1. b (2)
2. d (2)
3. c (3)
4. a (3)
5. b (4)
6. d (6)
7. c (6)
8. c (7)
9. e (9-11)
10. a (8)
11. d (8)
12. b (3, 11)
13. e (10-12)
14. a (10-12)
15. b (13)
16. c (13)
17. b (13)
18. c (14)
19. c (14)
20. a (14)
21. e (14)
22. e (15)
23. b (16)
24. a (16-17)
25. c (18)
26. b (19)
27. e (18-19)
28. c (20)
29. b (21-22)
30. c (22-23)

True/False

1. T (2)
2. F (6)
3. T (7)
4. F (10-12)
5. F (13)
6. T (13-14)
7. T (3,15-16)
8. F (18)
9. T (19-20)
10. T (21)

Short Answer

1. (2-3)
2. (3-5)
3. (9-12)
4. (12)
5. (13-14)
6. (15-16)
7. (17)
8. (19-20)
9. (20-21)
10. (22)

CHAPTER 2

CONCEPTS FOR SOCIAL AND CULTURAL THEORIES

Extended Abstract

A distinct change in the pattern of U.S. immigration was noted at the turn of the nineteenth century. Immigration from eastern and southern Europe outpaced immigration from northern and western Europe. The "new immigrants" included large numbers of Jews and Italians, and it was feared that these groups would not assimilate. However, data from the 1911 Immigration Commission Report revealed that the average weekly income of Jewish men was higher than that for white men while Italian men earned significantly less. How could this be? Stark maintains that social and cultural theories provide a framework for answering this question.

Societies vary in terms of social structure, stratification and social networks. A social structure is any property of a group such as population density. Since rewards are not evenly distributed among a society's members, societies may be stratified or "layered" by social class. Over time individuals or groups may experience upward or downward mobility, and social position or status may be based on merit (achieved status) or inherited (ascribed status). A complex web of social ties or networks also characterizes societies. These networks reflect varying degrees of power. Granovetter argues that weak, nonredundant ties enable persons to spread information effectively while strong, redundant networks facilitate influence. Similarly, local networks are small, more intimate and geographically concentrated; whereas, cosmopolitan networks are large, less intimate and geographically diverse.

Stark next notes that cultural and structural dimensions of social life must be explored. Key cultural concepts introduced in this chapter are values and norms, roles, multiculturalism and subcultures, prejudice and discrimination, assimilation and accommodation and modernization and globalization. Culture includes the things people learn and pass from generation to generation. Two important things learned are a group's values, ideas or beliefs about what is good or bad, and its norms, rules of acceptable and unacceptable behavior. Norms are combined to create social roles or behavioral guidelines that are associated with the various social positions we occupy like college student.

A society may be comprised of many different cultural groups. Multicultural societies are characterized by extensive cultural diversity. Particular racial and ethnic groups such as African Americans and Italians are examples of subcultures within U.S. society. However, within a multicultural society, one subculture may impose its cultural standards on others. When this occurs, multiculturalism may reinforce ascribed statuses and promote negative beliefs (prejudice) and encourage unequal treatment of groups (discrimination). Multiculturalism also raises questions concerning cultural assimilation and accommodation. If groups are expected to conform to prevailing dominant group norms, assimilation may occur. If groups seek to address shared interests, accommodation may occur. Globalization of culture is introduced also as an outcome of modernization, a process of extensive technological and economic change.

Stark then proceeds to apply these social and cultural concepts to the study of Jewish and Italian immigration to North America. Ethnic stereotypes were promoted in many publications at the beginning of the twentieth century, and by 1921, strict immigration quotas were established. Jews and Italians experienced less prejudice and discrimination as they began to assimilate North American culture, achieve economic success and as religious differences were accommodated. But, even with these changes, Jews experienced upward mobility sooner. Why?

Cultural theories suggest Jews experienced more mobility because educational achievement was stressed. Zborowski and Herzog note that traditional Jewish city (*shtetl*) life favored the maintenance of strong school systems and that Jewish families stressed the virtues of a good education. Consequently, Jewish immigrants were prepared to fill well-paying occupations as they immigrated to North America. Conversely, Covello observed that rural families from southern Italy stressed family loyalty over academic performance, and it was not unusual for children to quit school at an early age.

Developing a social theory of ethnic mobility, Steinberg maintains that an immigrant group's status in the home country is the best predictor of the group's eventual status in the country of destination. Steinberg observed that Jewish immigrants were previously concentrated in high status occupations while Italians were concentrated in lower status jobs. Analyzing data on educational attainment, Perlmann and later Glazer concurred that Jewish upward mobility was influenced by favorable occupational status in the home country and a value system that stressed education.

Turning to Italian immigration, Stark shows how reference group theory helps explain differences in ethnic mobility. For many first-generation Italian immigrants, southern Italy remained the primary reference group. Italian immigrants believed they would return home soon and were less willing to assimilate North American culture. In the final study, Greeley demonstrates that the importance of family loyalty among Italians has persisted. However, education may now be perceived as a viable means of honoring family obligations.

Key Learning Objectives

After a thorough reading of Chapter 2, you should be able to:

1. define society and culture and note how the concepts are interdependent.
2. distinguish key concepts associated with social and cultural theories.
3. understand how social and cultural theories are utilized to explain differences in ethnic mobility.
4. identify key characteristics of social structure.
5. state how mobility, status and class reflect stratification.
6. determine how network structures facilitate the spread of information and shape power relationships.
7. describe how values, norms and roles define and regulate social behavior.
8. identify the interdependent associations among multiculturalism, subcultural variation, ascribed status, prejudice, discrimination, assimilation and accommodation.

9. note how modernization impacts globalization.

10. identify examples of prejudice, discrimination, assimilation and accommodation experienced by Jewish and Italian immigrants to North America.

11. understand the different roles education plays in cultural explanations of ethnic mobility.

12. know how social position and educational attainment interact to enhance ethnic mobility.

13. determine how ethnic mobility may be tied to reference group identification.

14. document how traditional Italian cultural values have persisted throughout the twentieth century.

15. gain a clearer understanding of how data may be utilized in testing theoretical claims.

Chapter Outline

I. Introduction to Social and Cultural Analysis
   A. Changing Immigration Pattern
   B. Immigration Commission Report of 1911
   C. Economic Success Differences among Ethnic Groups

II. Concepts for Social Theories
   A. Society: A Web of Social Relationships
   B. Social Structure: Any Group Property
   C. Stratification: Unequal Distribution of Rewards among Groups.
      1. Groups may be organized into a class system.
      2. Groups may change social position through upward or downward mobility.
      3. Individuals or groups are identified according to their social position or rank.
      4. Rank may be based on merit (achieved status) or inheritance (ascribed status).
   D. Network Structures: Systems of Social Ties
      1. Weak, nonredundant ties spread information.
      2. Strong, redundant ties enhance influence.
      3. Local networks are small and geographically concentrated.
      4. Cosmopolitan networks are large and geographically scattered.

III. Concepts for Cultural Theories
   A. Culture: Things Each Generation Creates, Learns and Uses
   B. Values: Standards for Determining What Is Good and Bad

C. Norms: Rules that Guide Behavior
D. Roles: Collections of Norms Associated with Particular Positions
E. Multiculturalism and Subcultures: Cultural Pluralism and Cultural Distinctiveness
F. Prejudice and Discrimination: Negative Attitudes and Actions that Promote Inequality
G. Assimilation and Accommodation: Models of Cultural Conformity and Cultural Diversity
H. Modernization and Globalization: The Spread of Technological and Economic Change throughout the World

IV. Developing Cultural and Social Theories of Ethnic Mobility
A. The North American Experience: Jewish and Italian Immigrants
1. Urban settlement on East and West Coast of the U.S. and Canada.
2. Proliferation of ethnic stereotypes in the mass media.
3. Patterns of Jewish and Italian assimilation and accommodation.
4. Jewish immigrants experience upward mobility earlier.
B. Cultural Theories: Addressing Differences in Jewish and Italian Mobility
1. Zborowski and Herzog argue that the key is the traditional Jewish family's focus on the virtues of education.
2. Covello notes how family loyalty takes precedence over academic study among Italians from southern Italy.
C. Steinberg's Social Theory of Ethnic Mobility
1. A group's social status in the home country impacts its eventual social status in the country of destination.
2. Jewish immigrants were previously concentrated in high status occupations; whereas, Italians were concentrated in lower status jobs.
D. Perlmann and Glazer: Ethnic Mobility Is Influenced by Social Position and Cultural Values
E. Reference Groups and Cultural Assimilation
1. Southern Italy remained the reference group for Italian immigrants.
2. Italians were more reluctant to learn English.
3. North American migration was viewed as temporary.
4. Giannini's Bank of Italy and shifting reference group identity.
5. Greeley documents the persistence of Italian family solidarity and loyalty in contemporary North America.

Key Terms

Based on your reading of Chapter 2, you should be able to define and illustrate the key sociological concepts listed below. Page numbers are provided in parentheses as reference points.

Society (33)
Social Structure (33)
Stratification (35)
Classes (35)

Upward Mobility (35)
Downward Mobility (35)
Status (35)
Achieved Status (35)
Ascribed Status (35)
Caste (35)
Network Structures (35)
The Strength of Weak Ties (35)
Tie (35)
Redundant Tie (36)
Nonredundant Tie (36)
Structural Holes (36)
Bridge Ties (36)
Bridge Position (36)
Local Networks (37)
Cosmopolitan Networks (37)
Culture (39)
Values (39)
Norms (40)
Role (41)
Multiculturalism (41)
Subculture (41)
Prejudice (42)
Discrimination (42)
Assimilation (42)
Accommodation (43)
Modernization (43)
Globalization (43)
Anti-Semitism (46)
Shtetl Life (48-49)
Ghetto (49)
Hypotheses (55)
Reference Group (57)

Key Research Studies

Listed below are key research studies cited in Chapter 2. Familiarize yourself with the major finding of these studies. Page references are provided in parentheses.

Granovetter: weak network ties promote the spreading of information while strong network ties enhance influence (35-37).

DiMaggio and Lough: buyers prefer to purchase items from within their social network, but sellers prefer selling to people outside their social network (37).

Gans: neighborhoods with strong redundant social ties (urban villagers) are less able to block unfavorable urban policies (37-38).

Zborowski and Herzog: traditional urban Jewish families stressed the virtues of education and scholarship (48-50).

Covello: traditional Italian families from southern Italy stressed family loyalty and responsibility over academic learning (50-51).

Steinberg: among first-generation immigrants, a group's status in the society of destination is influenced by their status in their society of origin (52-55).

Perlmann (and Glazer): social position and cultural values impact ethnic mobility (55-57).

Greeley: family loyalty and solidarity persists among Italians in North America (59-61).

Info Trac Search Words

Enter these search terms to conduct more extensive investigations of key topics introduced in Chapter 2.

U. S. Immigration Laws

Population Density
Density is a property of social structure. Select one of the articles on density. See if you can determine how density impacts quality of life.

Subculture

Prejudice
Subdivisions: Analysis

Assimilation (Sociology)
Subdivisions: Research
Choose one of the research studies listed and note how assimilation is being measured and tested. What are the major findings of your chosen study?

Multiple Choice

Answers and page references are provided at the chapter end.

1. A major source of statistical data on the "new" immigration at the beginning of the twentieth century is the
   a. 1920 *General Social Survey.*
   b. 1850 *U.S. Census.*
   c. *Immigration Commission Report of 1911.*
   d. 1900 *Gallup Poll.*
   e. *International Social Survey Program* of 1910.

2. According to the *Immigration Commission Report of 1911*, average weekly income was lowest for __________ males 18 and over.
   a. Italian
   b. Irish
   c. Jewish
   d. white
   e. African American

3. According to Stark __________ is primarily an example of a social concept rather than a cultural concept.
   a. stratification
   b. assimilation
   c. accommodation
   d. a value
   e. a norm

4. Gender is primarily an example of
   a. population density.
   b. an ascribed status.
   c. an achieved status.
   d. a local network
   e. a cosmopolitan network.

5. If a physician's son becomes a nurse, the son has experienced __________ mobility.
   a. local
   b. upward
   c. cosmopolitan
   d. downward
   e. achieved

6. Race and ethnicity are examples of
   a. achieved status.
   b. role status.
   c. norm status.
   d. value status.
   e. ascribed status.

7. Granovetter observed that
   a. strong ties are more effective in exerting influence.
   b. weak ties are more effective in spreading information.
   c. strong ties are more effective in spreading information.
   d. a and b
   e. a and c

8. Which concept is used to describe network structures?
   a. redundant ties
   b. structural holes
   c. bridge ties
   d. bridge position
   e. all of the above

9. Cosmopolitan networks tend to be characterized by
   a. weaker ties and are scattered geographically.
   b. weaker ties and are geographically concentrated.
   c. strong ties and are geographically concentrated.
   d. strong ties and are scattered geographically.
   e. none of the above

10. Each of the following is a characteristic of local networks except
   a. dense networks.
   b. full of holes.
   c. person-to-person interaction.
   d. strong ties.
   e. Members are clustered geographically.

11. French-speaking and English-speaking Canada represent a society that
   a. is multicultural.
   b. is characterized by parallel networks.
   c. is characterized by separate networks based on language.
   d. includes distinct subcultures.
   e. all of the above

12. According to Stark, culture
   a. is measured by the sex ratio.
   b. is the organization of society on the basis of property, power and prestige.
   c. includes everything people learn and learn to use.
   d. is a system of nonredundant ties that confer power.
   e. consists of self-sufficient groups of people united by social relationships.

13. Telling the truth is an example of
   a. prejudice.
   b. discrimination.
   c. stratification.
   d. a value.
   e. a role.

14. Rules of acceptable and unacceptable behavior are examples of
   a. norms.
   b. statuses.
   c. roles.
   d. values.
   e. assimilation.

15. Conformity to norms often brings
   a. disapproval and punishment.
   b. punishment and approval.
   c. disapproval and rewards.
   d. rewards and punishment.
   e. approval and rewards.

16. Within the work environment, employer and employee are examples of
   a. norms.
   b. values.
   c. roles.
   d. prejudice.
   e. discrimination.

17. The Amish are a group characterized by a distinctive set of beliefs and customs. The Amish are a/an
   a. upwardly mobile group.
   b. downwardly mobile group.
   c. norm.
   d. subculture.
   e. value.

18. Stark maintains that prejudice and discrimination are generally rooted in
   a. achieved status.
   b. ascribed status.
   c. upward mobility.
   d. downward mobility.
   e. none of the above

19. Negative attitudes about individuals or groups are often viewed as examples of
   a. multiculturalism.
   b. assimilation.
   c. accommodation.
   d. prejudice.
   e. discrimination.

20. Begonians are the dominant group in Doberia and Sardonians are expected to conform to the expectations of the Begonians. This is an example of
   a. globalization.
   b. modernization.
   c. assimilation.
   d. accommodation.
   e. none of the above

21. Stark argues that globalization has been spread primarily through
  a. modernization.
  b. caste systems.
  c. prejudice.
  d. discrimination.
  e. class systems.

22. Jewish and Italian immigrants to the U.S. settled primarily in the __________ because this region offered the greatest economic opportunity.
  a. rural South
  b. urban East
  c. Midwest
  d. urban South
  e. rural East

23. Prejudice and discrimination against Jews is known as
  a. globalization
  b. multiculturalism.
  c. local networks.
  d. cosmopolitan networks.
  e. anti-Semitism.

24. Stark sees intermarriage as a major outcome of
  a. prejudice.
  b. discrimination.
  c. globalization.
  d. assimilation.
  e. none of the above

25. Zborowski and Herzog discovered that schools were easily maintained in Jewish communities because
  a. Jews lived primarily in towns and cities.
  b. Jews lived primarily in rural areas.
  c. Jewish families practiced home schooling.
  d. Jewish schools were located in synagogues.
  e. none of the above

26. Which of the following statements about Italian immigrants to North America is true?
  a. Italian children dropped out of school at an early age.
  b. The majority of Italian immigrants came from southern Italy.
  c. Academic learning was not a strong value in southern Italian culture.
  d. Traditional Italian families viewed schooling as threatening a family's well-being.
  e. Each statement is true.

27. Steinberg argued that Jewish immigrants were more upwardly mobile than Italian immigrants because in the home country
    a. Jews were heavily concentrated in higher-status occupations.
    b. Jews were heavily concentrated in low-status occupations.
    c. Jews experienced more prejudice and discrimination.
    d. Italians experienced less prejudice and discrimination.
    e. Italians placed more emphasis on education.

28. Perlmann's research indicates that the best explanation for the Jewish-Italian difference in ethnic mobility is offered by
    a. the social theory.
    b. the cultural theory.
    c. the economic theory.
    d. both the social and the cultural theory.
    e. both the cultural and the economic theory.

29. The reference group for many first-generation Italian immigrants was
    a. western Europe.
    b. southern Italy.
    c. Canada.
    d. *shtetl* life in Poland.
    e. the U.S.

30. Utilizing data from the *International Social Survey Program*, Greeley observed that
    a. Canada is the primary reference group for Italian immigrants.
    b. shtetl life in Italy lead Italian families to stress the virtues of education.
    c. Italians still stress family loyalty and solidarity in contemporary U.S. society.
    d. since Italian immigrants were previously employed in high status jobs, they assimilated U.S. culture faster than any other immigrant group.
    e. all of the above

True/False

Answers and page references are provided at the chapter end.

1. At the beginning of the twentieth century, the "new" immigrants to the U.S. were primarily from southern and eastern Europe.

2. Compared to Jewish immigrants, Italian immigrant groups achieved greater economic success in North America earlier.

3. Population density, the sex ratio and a population's age composition are examples of social structure.

4. Stratification portrays the equal distribution of rewards and opportunities among different groups within a society.

5. Strong network ties are more effective in spreading information.

6. Cosmopolitan networks are concentrated geographically; whereas, local networks are scattered geographically.

7. Values are ideals or ideas about what a society judges as desirable or undesirable.

8. Discrimination deals primarily with negative attitudes and hostile beliefs while prejudice focuses on actions that promote inequality.

9. The U.S. imposed strict immigration quotas on persons of different nationalities from 1921 to 1965.

10. Steinberg noted that first-generation Jewish immigrants were employed primarily in lower status jobs prior to immigrating to North America.

Short Answer Questions

These short answer questions are provided to test your knowledge and understanding of the basic sociological concepts presented in Chapter 2. Page references for answers are included at the chapter end.

1. A change in U.S. immigration patterns was observed at the beginning of the twentieth century. What were some of the major differences between the "new" and "old" immigrants?

2. How are the concepts, society and culture, distinct but related?

3. What criteria are used to determine whether a status is achieved or ascribed?

4. Why do cosmopolitan networks provide groups with a low level of solidarity?

5. What are some of the possible outcomes of multiculturalism?

6. Identify several of the distinguishing features of subcultures.

7. Assimilation and accommodation are patterns of cultural interaction. How do they differ?

8. Identify examples of prejudice and discrimination experienced by early Jewish and Italian immigrants to North America.

9. The studies on ethnic mobility by Zborowski and Herzog and Covello suggest that Jewish and Italian immigrants were influenced by different cultural factors. What were these factors?

10. According to Steinberg, what is the nature of the association between immigration and social status?

Essay Questions

These questions are designed to test your understanding of key sociological concepts presented in Chapter 2 and your ability to apply these insights to concrete situations.

1. Do ascribed or achieved status or both primarily influence mobility in contemporary U.S. society? Provide illustrations to justify your position.

2. How is the Internet redefining our understanding of local and cosmopolitan networks?

3. Identify three important values shared by U.S. residents. How might these values vary by geographic region and place of residence?

4. Does contemporary U.S. society primarily stress cultural assimilation or accommodation? Provide concrete examples to justify your argument.

5. Apply Steinberg's theory of ethnic mobility to the current Japanese and Mexican immigration to the U.S. Does the theory hold for these groups?

Answers

Multiple Choice

1. c (32-33)
2. a (33)
3. a (33-35)
4. b (35)
5. d (35)
6. e (35)
7. d. (35-36)
8. e (36)
9. a (37)
10. b (37)
11. e (38-43)
12. c (39)
13. d (39)
14. a (40)
15. e (40)
16. c (41)
17. d (41)
18. b (42)
19. d (42)
20. c (42)
21. a (43)
22. b (44)
23. e (46)
24. d (47)
25. a (48-49)
26. e (50-51)
27. a (52-55)
28. d (55-57)
29. b (58)
30. c (59-61)

True/False

1. T (31)
2. F (32-33)
3. T (33-34)
4. F (35)
5. F (35)
6. F (37)
7. T (39)
8. F (42)
9. T (46-47)
10. F (53)

Short Answer

1. (31-33)
2. (33, 39)
3. (35)
4. (37-38)
5. (41-42)
6. (41-42)
7. (42-43)
8. (44-47)
9. (48-51)
10. (52-55)

CHAPTER 3

MICROSOCIOLOGY: TESTING INTERACTION THEORIES

Extended Abstract

Social and cultural forces that are external to the individual shape human behavior. Durkheim labeled these forces social facts. As persons begin to interact with others, they are presented with choices and begin to internalize the norms and the expectations of social groups like the family. This process is known as socialization, and it begins at birth. Microsociology is concerned with understanding how people make choices within a given socio-cultural environment. Microsociological theories are developed in the first half of this chapter while theory testing is explored in the second half.

Microsociologists maintain that choices are based on rational decisions rather than random events. According to the rational choice proposition, human actions may be restricted, but they are based on preferences that are designed to maximize perceived benefits. Microsociological theories are concerned with understanding how humans perceive benefits and costs and how these benefits and costs are constructed and shared through social interaction. Two major interaction theories are presented. They are symbolic interaction theory and exchange theory.

Symbolic interaction theories attempt to understand human communication patterns, which are highly symbolic in nature. How do people perceive events? How do they gain a sense of self through interaction with others? In answering these questions, Stark provides concise introductions to the thought of three classic symbolic interactionists, Blumer, Cooley and Mead. Blumer argued that human actions are based on the perception of events. This perception is influenced by one's contact with others and is subject to modification. Likewise, Cooley argued that people are like mirrors, and self-images are reflected or exchanged through social interaction. This concept was known as the "looking glass self." Finally, Mead argued that the self emerges once persons are able to comprehend symbolic forms of communication and are able to perceive events as others may perceive them. Mead called this taking the role of the other.

Exchange theorists argue that, human beings act in ways that enable them to maximize perceived benefits and minimize perceived costs. Over time stable patterns of exchange develop among persons, and this enhances attachments and social solidarity. Homans argues that participation in common activities increases the likelihood that people will like each other (law of liking), will agree with each other (law of agreement) and will conform (law of conformity). Furthermore, as people interact, stronger attachments are formed among persons of equal rank (law of inequality). Stark concludes that group ties (solidarity) enhance group conformity (law of conformity) and empower a society's norms. When persons violate norms, attachments are risked.

Stark next turns to theory testing. Theories are abstractions, but sociologists evaluate theories by testing hypotheses and analyzing relationships among variables. A variable is a concept that measures change. Educational attainment is a variable. One may possess a high or a low degree of educational attainment. Variables may also influence each other. Independent variables are sometimes identified as "causes" and dependent variables are viewed as "consequences."

Cause and effect associations among variables (causality) are determined once time order and correlation are established and a test for spuriousness has been performed. Time order involves identifying the independent (cause) and dependent (effect) variables. Correlation is a measure of association. Correlation indicates that a change in one variable is associated with a change in another variable. Correlations may be strong or weak or positive or negative. A positive correlation indicates that both variables are moving in the same direction while a negative correlation means they are moving in opposite directions. A spurious relationship exists when two variables appear to be associated because they share common ties to a third variable. Spurious relationships challenge cause-and-effect associations.

Researchers may employ a variety of methods in testing theoretical claims. Two methods commonly used by microsociologists are experiments and field research. Asch's study of solidarity and conformity is offered as an example of experimental research design. With the experimental research design, researchers can randomly assign persons to different study groups and control which group will be introduced to the effects associated with the independent variable. In Asch's experiment, group solidarity was the independent variable and conformity was the dependent variable. Asch observed that statistically significant differences existed among student groups and that conformity was higher among unanimous rather than partial agreement groups. Stark maintains that experiments are powerful research tools because they enable researchers to establish time order, correlation and test spuriousness.

A study on new religious movement membership recruitment by Stark and Lofland is offered next as an example of field research. With field research the researcher is directly involved in studying human behavior as it is actually taking place. Lofland and Stark studied Unification Church recruitment patterns by observing the group's religious activities. Initially, Lofland and Stark were covert rather than overt observers. The group members were thus initially not aware that they were being studied. Lofland and Stark concluded that most persons joined the movement because strong friendship ties had been established with the group's members. Persons were not initially attracted by the group's beliefs. Stark argues that the Heaven's Gate movement failed because it was unable to recruit members through close friendship networks.

Stark concludes Chapter 3 by developing four principles of social movement recruitment. These principles are grounded in the interaction of strong and weak social networks. Social movements tend to originate among strong local networks while expansion is tied to weak cosmopolitan networks.

Key Learning Objectives

After a thorough reading of Chapter 3, you should be able to:

1. identify a social fact.
2. understand how behavior is influenced by benefits, costs and preferences.
3. compare and contrast two major interaction theories (symbolic interaction and exchange).

4. note the major contributions of Blumer, Cooley and Mead to the development of symbolic interaction theory.

5. comprehend the complex association among exchange, solidarity inequality, agreement and conformity.

6. conceptualize a variable.

7. state the difference between an independent and a dependent variable.

8. apply the three principles of causality within a theory testing context.

9. interpret a correlation coefficient.

10. clearly distinguish experimental research design and field research.

11. identify Asch's major findings on solidarity and conformity.

12. know why researchers employ tests of significance.

13. describe the relationship among ideology, social attachment and social movement membership recruitment.

14. clearly distinguish covert and overt observation techniques.

15. note how local and cosmopolitan networks enhance the development and growth of social movements.

Chapter Outline

I. Microsociology and Interaction Theories
   A. Social Facts, Cultural Forces and Choice
   B. The Rational Choice Proposition
      1. Choices are generally rational.
      2. Choices may be limited.
      3. Choices are guided by preferences.
      4. Humans tend to maximize perceived benefits.
   C. Symbolic Interaction Theory: Blumer, Cooley and Mead
      1. Blumer maintains that actions are grounded in perceptions of events and perceptions are influenced by social interaction.
      2. Cooley introduces the "looking glass self." People reflect perceived images.
      3. Mead observes that the sense of self emerges as persons are able to take the role of the other.
   D. Exchange Theory: Homans

1. The law of liking states that participation in common activities builds attachments.
2. Emotional attachments are strong among persons of equal rank.
3. Close-knit groups experience more conformity.

II. Principles of Theory Testing
- A. Variables versus Constructs
- B. Independent and Dependent Variables
  1. Independent variables are associated with "causes."
  2. Dependent variables are associated with "consequences."
- C. Three Criteria for Establishing Causality
  1. Time order involves identifying what comes first.
  2. Correlation measures the association among variables.
     - a. Correlations may be weak or strong.
     - b. Correlations may be positive (same direction variation) or negative (different direction variation).
  3. Spuriousness tests for the presence of correlations that are not causal.

III. Research Design and Data Collection
- A. Experimental Design and the Study of Group Solidarity
  1. Asch employs experimental research design to identify the association between group pressure and conformity.
  2. A researcher may manipulate the independent variable with experimental design.
  3. Comparison groups are assigned randomly.
  4. Tests of significance identify important research findings.
- B. Field Research and the Study of Recruitment
  1. Field research lets investigators observe behavior as it is actually taking place.
  2. Groups may be aware that they are being studied (overt observation) or unaware (covert observation).
  3. Lofland and Stark utilize field research techniques to study Unification Church recruitment methods.
  4. Lofland and Stark observe that friendship attachments are a strong recruitment (conversion) mechanism for the Unification Church.
  5. Heaven's Gate movement failed because strong network attachments among members and potential converts were not established.
  6. Stark develops four principles of social movement recruitment.
     - a. Open-minded principle: recruitment is not linked to movement's ideology.
     - b. Network recruitment principle: recruitment is linked to interpersonal attachments.
     - c. Cosmopolitan growth principle: organizational growth is tied to establishing strong nonredundant tie networks.
     - d. Principle of dense origins: movements begin where with local (redundant) networks are strong.

Key Terms

Based on your reading of Chapter 3, you should be able to define and illustrate the key sociological concepts listed below. Page numbers are provided in parentheses as reference points.

- Social Facts (67)
- Socialization (68)
- Internalized (68)
- Rational Choice Proposition (69)
- Altruism (70)
- Social Interaction (71)
- Symbolic Interaction (72)
- Symbols (72)
- Looking Glass Self (74)
- Mind (74)
- Self (74)
- Taking the Role of the Other (74)
- Exchange Theory (75)
- Law of Liking (76)
- Attachment (76)
- Law of Inequality (76)
- Law of Agreement (76)
- Law of Conformity (77)
- Constant (77)
- Variable (77)
- Cause (77)
- Independent variable (77-78)
- Dependent variable (77-78)
- Time Order (78)
- Correlation (78, 80)
- Spurious Relationship (78)
- Abstractions (78)
- Correlation Coefficient (80)
- Experiment (82)
- Subjects (82)
- Randomization (83)
- Test of Significance (83)
- Manipulate (83)
- Social Movement (83)
- Ideology (84)
- Field Research (84)
- Covert Observers (85)
- Observer Effects (85)
- Overt Observers (85)
- Replication (86)
- Open-minded Principle (87)
- Network Recruitment Principle (87)
- Cosmopolitan Growth Principle (88)
- Principle of Dense Origins (88)

Key Research Studies

Listed below are key research studies cited in Chapter 3. Familiarize yourself with the major findings of these studies. Page references are provided in parentheses.

Blumer: actions are based on perceptions that arise from and are modified by social interactions (73-74).

Homans: developed laws of exchange involving liking, weak emotional attachments and unequal rank, agreement, group solidarity and conformity (76-77).

Asch: conformity increases as group pressure increases (81-82).

Lofland and Stark: strong within group attachments enhance conversion to new religious movements (84-87).

Stark and Bainbridge: new religious movements recruit members from among the religious inactive and dissatisfied (87).

Stark: social movements originate among strong, local networks while growth is tied to weak, cosmopolitan networks (87-88).

Info Trac Search Words

Enter these search terms to conduct more extensive investigations of key topics introduced in Chapter 3.

Rational Choice Theory
Select one of the studies listed. How is rational choice theory being used to test the phenomenon studied?

Symbolic Interactionism

Correlation (Statistics)

Spuriousness
Take a look at one of the studies cited. How is spuriousness demonstrated?

Peer Pressure
Subdivisions: Social Aspects

Multiple Choice

Answers and page references are provided at the chapter end.

1. As people internalize the expectations of family life, they experience
    a. individualization.
    b. socialization.
    c. tests of significance.
    d. replication.
    e. spuriousness.

2. According to the rational choice proposition, choices are
    a. essentially irrational.
    b. a product of inheritance.
    c. grounded in personal preferences and tastes.
    d. primarily impulsive.
    e. genetically determined.

3. Microsociology attempts to understand
    a. social interaction.
    b. human exchanges.
    c. patterns of interaction and exchange.
    d. all of the above
    e. none of the above

4. Compared to exchange theory, symbolic interaction theory is concerned more with
    a. how people develop a sense of self.
    b. how social solidarity is maintained.
    c. identifying spurious relationships.
    d. covert rather than overt observation.
    e. tests of significance.

5. Blumer maintained that
    a. the real world is clearly understood by all.
    b. communication is highly symbolic.
    c. human behavior is irrational and selfish.
    d. one's perceptions influences one's view of the world.
    e. b and d

6. A person may hesitate to take the last sandwich on a plate because that person does not want to be perceived as being "selfish." This scenario illustrates
    a. Cooley's looking glass self.
    b. Homans' law of inequality.
    c. Stark's open-minded principle.
    d. the three criteria of causality.
    e. experimental research design.

7. Looking at Mead's understanding of symbolic interaction, the mind is to the self as
    a. rational is to irrational.
    b. tastes are to preferences.
    c. understanding symbols is to taking the role of the other.
    d. costs are to benefits.
    e. covert actions are to overt actions.

8. Exchange theory states that
    a. people seek to maximize rewards.
    b. people seek to maximize costs.
    c. exchanges are limited by self-interests.
    d. a and c
    e. b and c

9. Stable bonds of affection among people are known as
   a. symbols.
   b. attachments.
   c. correlations.
   d. internalizations.
   e. constants.

10. Persons who like each other are more likely to agree with each other according to according to Homans' law of
   a. agreement.
   b. inequality.
   c. conformity.
   d. group pressure.
   e. likely.

11. Male and female voting patterns may differ substantially. A researcher studying this difference will treat gender as a
   a. communication cue.
   b. spurious relation.
   c. symbol.
   d. variable.
   e. correlation.

12. A study indicated that economic success is influenced strongly by age. Age would be
   a. the independent variable.
   b. the dependent variable.
   c. the constant.
   d. a positive correlation.
   e. a negative correlation.

13. In establishing time order, the
   a. effect comes before the cause.
   b. cause comes before the effect.
   c. independent variable influences the dependent variable.
   d. a and c
   e. b and c

14. The three criteria of causation are
   a. constants, variables and spurious relationships.
   b. symbols, mind and self.
   c. attachments, solidarity and group pressure.
   d. time order, correlation and nonspuriousness.
   e. overt, covert and internal measures.

15. Correlations may be
  a. positive.
  b. negative.
  c. weak.
  d. strong.
  e. all of the above

16. Variables that appear to be correlated because they share an association with a common third variable are known as __________ relationships.
  a. independent
  b. dependent
  c. spurious
  d. positive
  e. negative

17. Data from the *Nations of the Globe* data source indicate that the fertility rate declines as economic development increases. This is an example of a
  a. positive constant.
  b. negative constant.
  c. neutral correlation.
  d. negative correlation.
  e. positive correlation.

18. A theory is an example of a/an
  a. correlation.
  b. abstraction.
  c. spurious association.
  d. constant.
  e. variable.

19. Asch's experiment indicates that group pressure influences
  a. conformity.
  b. inequality.
  c. cosmopolitan networks.
  d. recruitment.
  e. local networks.

20. Asch flipped a coin to determine whether the student respondent would participate with a unanimous group or a group reflecting partial solidarity. This assignment process followed a procedure known as
  a. test of significance.
  b. replication.
  c. randomization.
  d. internalization.
  e. socialization.

21. Asch observed that conformity increased as group solidarity increased. This finding is an example of a/an __________ correlation.
    a. spurious
    b. independent
    c. dependent
    d. negative
    e. positive

22. Persons who participate in an experiment are known as
    a. recruits.
    b. subjects.
    c. field workers.
    d. constants.
    e. experimenters.

23. Researchers employ tests of significance to determine whether an observed correlation is due to a chance or random variation. Researchers like to limit the odds of a chance finding to
    a. 5 to 1.
    b. 10 to 1.
    c. 20 to 1.
    d. 100 to 1.
    e. either c or d

24. Civil Rights organizations emerged during the 1950s and 1960s as a concrete way to address problems associated with racial inequality. These groups were part of a/an
    a. random fact.
    b. covert operation.
    c. social movement.
    d. abstraction.
    e. spurious network.

25. The Lofland and Stark study on new religious movements links conversion to
    a. ideology.
    b. symbols.
    c. attachments.
    d. a quest for meaning.
    e. isolation.

26. When a researcher is observing behavior as it actually takes place within its natural environment, data are being collected utilizing this research design format.
    a. field research
    b. test of significance
    c. manipulation
    d. an experiment
    e. randomization

27. A researcher joins a car theft ring to gain information on how the group steals cars. The ring members do not know that their new member is a researcher. This is an example of
    a. overt observation.
    b. spurious observation.
    c. random observation.
    d. covert observation.
    e. experimental observation.

28. Stark has observed that converts to new religious movements are recruited from
    a. strong political backgrounds.
    b. weak political backgrounds.
    c. strong religious backgrounds.
    d. weak religious or irreligious backgrounds.
    e. any religious or political background.

29. The Heaven's Gate movement failed because
    a. the group placed too much emphasis on ideology as a recruitment method.
    b. people stopped believing in UFOs.
    c. strong attachments were formed between group members and recruits.
    d. the group only recruited men.
    e. the leader of the group remarried.

30. Stark postulates that social movements originate within dense __________ networks while growth opportunities are limited to __________ networks.
    a. cosmopolitan; local
    b. local; cosmopolitan
    c. independent; dependent
    d. dependent; independent
    e. overt; covert

True/False

Answers and page numbers are provided at the chapter end.

1. According to the rational choice proposition, persons tend to maximize perceived rewards and minimize perceived costs.

2. The exchange of rewards is a basic social interaction.

3. Cooley's "looking glass self" illustrates how the sense of self is individually constructed.

4. Mead associated the ability to comprehend symbolic communication with the development of the self.

5. According to Homans' law of inequality, emotional attachments are stronger among persons of the same rank.

6. Independent variables are to dependent variables as "causes" are to consequences."

7. When a positive correlation is observed between two variables, the variables are changing in the opposite direction.

8. Researchers employ tests of significance to see if the changes observed among variables are meaningful or due to chance.

9. Field research involves observing behavior in a controlled, laboratory setting.

10. Stark's open-minded principle states that social movements originate among dense cosmopolitan networks.

Short Answer Questions

These short answer questions are provided to test your knowledge and understanding of the basic concepts presented in Chapter 3. Page references for answers are provided at the chapter end.

1. What is a social fact, and how do social facts impact human choices?

2. How might altruism illustrate the rational choice proposition?

3. Compare and contrast symbolic interaction theory and exchange theory.

4. Identify Blumer's three premises of symbolic interaction.

5. Distinguish Homans' law of liking, inequality, agreement and conformity?

6. What do positive and negative correlations mean with respect to the observed association between or among variables?

7. Identify the three criteria of causation.

8. According to Asch's experiment, how does group pressure impact individual conformity?

9. What did Lofland and Stark learn about the association among social attachments, ideology and conversion to new religious movements?

10. Distinguish covert and overt observation techniques as they relate to field research.

Essay Questions

These questions are designed to test your understanding of key sociological concepts presented in Chapter 3 and your ability to apply those insights to concrete situations.

1. A person is presented with two job offers. According to the rational choice proposition, what factors could determine which offer is ultimately selected?

2. Design a research project and specify your independent and dependent variable. Explain how you will obtain and evaluate your data and how you will test for spuriousness.

3. A researcher wants to know if the use of a study guide will improve student test performance in introductory sociology classes at a particular college. How would the researcher conduct this study employing an experimental research design?

4. What ethical issues may be encountered when a researcher employs covert observation techniques in gathering data?

5. A local community service organization has been loosing members for years. How might an understanding of local and cosmopolitan networks enable the organizations leadership to recruit more members?

Answers

Multiple Choice

1. b (68)
2. c (69)
3. d (71-72)
4. a (71, 74)
5. e (73-74)
6. a (74)
7. c (74)
8. d (75)
9. b (76)
10. a (76-77)
11. d (77)
12. a (77-78)
13. e (78)
14. d (78)
15. e (78, 80)
16. c (78)
17. d (80)
18. b (78-79)
19. a (81-82)
20. c (83)
21. e (78, 80, 82-83)
22. b (82)
23. e (83)
24. c (83)
25. c (84-86)
26. a (84)
27. d (85)
28. d (87)
29. a (88-89)
30. b (88)

True/False

1. T (69-70)
2. T (71)
3. F (74)
4. F (74)
5. T (76)
6. T (77-78)
7. F (78, 80)
8. T (83)
9. F (84)
10. F (87-88)

Short Answer

1. (67-69)
2. (69-71)
3. (72-77)
4. (73-74)
5. (76-77)
6. (78, 80)
7. (78-79)
8. (81-83)
9. (84-87)
10. (85)

CHAPTER 4

MACROSOCIOLOGY:
STUDYING LARGER GROUPS AND SOCIETIES

Extended Abstract

Sociologists utilize different techniques to study large group behavior. In this chapter readers are introduced to research methods and theories that help provide a better understanding of large group behavior. New methodological approaches introduced in this chapter include survey research and comparative sociology. Functionalism, social evolution and conflict theory are presented as examples of macrosociological theories.

Aggregate data (proportional facts) on populations are obtained from censuses and samples. A census can be very expensive because it is based on a count of an entire population. Consequently, researchers rely on representative samples. Sample accuracy is improved when random selection techniques are employed. This approach assures the researcher that each case in the larger population has a more equal chance of being included in the sample. A popular random selection technique is random digit dialing.

Stark next demonstrates how experimental design, survey research and network analysis can be employed to study larger social structures. A classic experimental study by Darley and Latane' illustrates how perceived group size affects a person's willingness to help someone experiencing distress. Generally, individuals in larger groups are less likely to provide help.

The study on religion and delinquency by Hirschi and Stark is an example of survey research and is based on a sample of junior high and high school students from a medium size Californian city. Initially, they observe that frequent church attendees are less delinquent. However, when they looked at the gender specific effects, they discover that the relationship is spurious; the correlation between church attendance and delinquency disappears. Subsequent research on religion and delinquency conducted by researchers using samples from different parts of the U.S. yields conflicting findings. This leads Stark to conclude that the variations in the findings are due to differences in social surroundings or contextual effects. Therefore, the impact of religion on delinquency is limited to areas where church membership rates are high.

Networks are a social structure characteristic and become the unit of analysis when network analysis is employed as a research method. Zablocki employs network analysis in his study of love relationships in communal groups in the U.S. The intensity of love relationships in various communal groups is gauged, and it is hypothesized that communal groups characterized by more extensive mutual love relationships are more stable. The opposite effect is observed. Membership turnover and group disintegration is higher among groups characterized by more extensive love relationships because these relationships promote jealousy among members.

Turning to the study of societies, Stark argues that societies are systems comprised of interdependent parts that maintain stability and equilibrium. An example of a part is a social institution like the family or religion. Institutions are a cluster of groups and roles designed to address a particular need. Institutions are interdependent, and they adapt to change in ways that maintain stable relationships. However, sociologists do not agree on the nature of this interdependence or how institutions respond to change and maintain stability. Three macrosociological theoretical paradigms are functionalism, social evolution and the conflict perspective.

Functionalism assumes that each part in a system serves a purpose. When the part functions properly, potential disruptions to the system are stabilized. Thus, extended families exist in many rural, agricultural societies in order to provide care for the young, the old and the disabled. In urban, industrial societies, this care may be provided through retirement benefits. These benefits are a functional equivalent. Social evolutionary theories place more emphasis on how groups or institutions adapt to changing social environments like increasing urbanization. Successful adjustments to changing social conditions enhance a group's or institution's ability to remain viable and survive. Conflict theories explain how power relationships structure society. Two classical conflict theorists are Marx and Weber. Marx argued that a society's dominant class constructs a society that will protect its own position and interests. On the other hand, Weber argued that conflict exists among common interest or status groups such as racial, ethnic, gender or religious groups. Differences in power between and within status groups create systems of inequality in a given society.

Stark concludes Chapter 4 by providing an introduction to comparative sociological research. Here the society is the unit of analysis. Major sources of data for comparative research among non-industrial societies are the *Standard Cross-Cultural Sample* and the *Atlas of World Cultures*. Chagnon's study on violence among the Yanomamo and Paige's study of conflict among non-industrial societies are offered as examples of comparative sociological research. Chagnon's field research on the Yanomamo is included in the *Standard Cross-Cultural Sample*. His studies indicate that revenge killings are common among the Yanomamo. Analyzing data from the *Atlas of World Cultures*, Paige observed that violence is higher in patrilocal rather than matrilocal societies. In matrilocal societies males are more likely to interact with males that are not kin since a married couple resides near the bride's family. This reduces the creation of factional groups, which in turn reduces violence.

Key Learning Objectives

After a thorough reading of Chapter 4, you should be able to:

1. provide examples of simple and proportional facts.
2. distinguish populations and samples.
3. state how random selection improves the validity of samples.
4. describe the association between group size and individual responsibility.
5. apply the three criteria for causality to the analysis of data obtained from survey research.
6. know how to test for contextual effects.
7. identify the major components of survey research.

8. apply network analysis to the comparative study of groups of varying size.

9. explain how society may be perceived as a complex system of interdependent social institutions.

10. identify the major theoretical tenets of functionalism, social evolutionary theory and conflict theory.

11. provide examples of functional alternatives and dysfunctions.

12. state some of the major criticisms of the social evolutionary approach.

13. compare and contrast the Marxian and Weberian approach to the study of conflict.

14. identify major cross-cultural sources of data on non-industrial societies.

15. specify the interrelationship among kinship, residence patterns, the creation of factional groups and violence.

Chapter Outline

I. Macrosociological Research
  A. Simple and Proportional Facts: Different Types of Data
    1. Simple facts are based on a small number of observations.
    2. Proportional facts are based on sample data.
  B. Populations, Censuses and Samples
    1. Censuses are comprehensive but expensive.
    2. Samples based on random selection are more likely to reflect population characteristics.
  C. Case Studies
    1. Darley and Latane''s experimental study indicates that group size influences an individual's willingness to assume responsibility.
    2. Hirschi and Stark discover that contextual effects monitor the association between church membership and delinquency.
      a. Initially the relationship between church membership and delinquency was believed to be spurious.
      b. The impact of religion on delinquency is important in areas where church membership is high.
    3. Zablocki applies network analysis to the study of love and jealousy in communal settings.
      a. Hypothesized that strong love networks would increase group solidarity.
      b. Strong love networks promote jealousy and erode solidarity.

II. Societies as Social Systems
  A. Institutions as Vital Social Parts
  B. The Interdependent Nature of Social Institutions

C. Effective Institutions Promote Stability and Equilibrium

III. Major Macrosociological Theoretical Perspectives
   A. Functionalism: the Intent and Purpose of Parts
      1. Effective parts preserve order and combat potential disruption.
      2. Some parts are interchangeable (functional alternatives).
      3. Some social structures disrupt society's equilibrium (dysfunction).
   B. Social Evolutionary Theories: Change and Adaptation
      1. Institutions that are able to adapt to change tend to survive.
      2. Societies have become more complex and urban over time.
      3. Change is neither inevitable nor a sign of progress.
   C. Conflict Theory: Power Relationships Structure Society
      1. Marx maintained that power struggles are grounded in class conflicts.
      2. Weber argued that conflict is rooted in competition among status groups.

IV. Comparative Sociological Research
   A. Societies as the Unit of Analysis
   B. Primary Data Sources for the Study of Non-Industrial Societies
      1. Murdock and White's *Standard Cross-Cultural Sample* represents a cross-section of 186 non-industrial societies.
      2. Murdock's *Atlas of World Cultures* represents a larger representative cross-section of 563 non-industrial societies.
   C. Comparative Research Case Studies
      1. Chagnon studies revenge violence among the Yanomamo and contributes data to the *Standard Cross-Cultural Sample.*
      2. Utilizing data from the *Atlas of World Cultures*, Paige links violence in non-industrial societies to kinship structure and matrilocal residence patterns.
         a. Matrilocal societies are more communal and experience less violence.
         b. Patrilocal societies are more factional and experience more violence.

Key Terms

Based on your reading of Chapter 4, you should be able to define and illustrate the key sociological concepts listed below. Page numbers are provided in parentheses as reference points.

Data Collection (96)
Simple Fact (96)
Proportional Fact (98)
Census (98)
Sample (98-99)
Random Selection (99)
Survey Research (100)

Contextual Effect (102)
Network Analysis (104)
Commune (104)
System (108)
Social Institutions (108-109)
Interdependence (108)
Equilibrium (109)

Open Systems (109)
Functionalist Theories; Functionalism (110)
Nuclear Family (110)
Extended Family (110)
Functional Alternative (111)
Dysfunctions (111)
Social Evolutionary Theories (111)
Involution (112)
Conflict Theory (112)
Status Group (113)
Comparative Research (114)
Standard Cross-Cultural Sample (116)
Atlas of World Cultures (116)
Factional Societies (117)
Communal Societies (117)
Kinship (117)
Residence (117)
Patrilocal Rule of Residence (118)
Matrilocal Rule of Residence (118)

Key Research Studies

Listed below are key research studies cited in Chapter 4. Familiarize yourself with the major findings of these studies. Page references are provided in parentheses.

Darley and Latane': willingness to assume personal responsibility in crisis situations varies by group size (97).

Hirschi and Stark: the statistical significance of the effect of church attendance on delinquency varies by region suggesting that contextual effects are present (100-103).

Zablocki: intense love networks encourage jealousy and endanger communal social stability (104-107).

Stinchcombe: identifies three major components of functionalist theories (110).

Chagnon: documents a high degree of revenge killing among Yanomamo factional groups (116-117).

Paige: links violence in non-industrial societies to kinship ties and residence patterns among communal and factional groups (117-119).

Info Trac Search Words

Enter these search terms to conduct more extensive investigations of key topics introduced in Chapter 4.

Statistical Sampling
Subdivisions: Technique

Social Institutions

Functionalism (Social Sciences)

Select one of the articles cited and note how the functionalist perspective provides an understanding of the topic being explored.

Max Weber (Sociologist)

Choose one of the articles listed. How are power relationships described in the chosen study?

Kinship

Subdivisions: Social Aspects

Multiple Choice

Answers and page references are provided at the chapter end.

1. Darley and Latane′ observed that an individual is more likely to respond in a crisis situation when
   a. the person needing help is female.
   b. the person needing help is male.
   c. The size of the group is large.
   d. The size of the group is small.
   e. Eye contact can be made with the victim.

2. Contemporary data on religious behavior indicate that females are more religious than males and that African Americans are more religious than whites. These are examples of __________ facts.
   a. simple
   b. proportional
   c. contextual
   d. random
   e. factional

3. The *World Values Survey* is an example of a/an
   a. experimental research design.
   b. census.
   c. population.
   d. theoretical paradigm.
   e. sample.

4. If a survey is based on random selection, this means that
   a. each person in the chosen population has an equal chance of being in the sample.
   b. the researcher's friends are more likely to be included in the sample.
   c. people will be asked to participate in the study only once.
   d. The sample will include only persons who have participated in previous studies.
   e. none of the above

5. Survey research relies heavily on data collected from
   a. personal interviews.
   b. questionnaires.
   c. samples.
   d. all of the above
   e. none of the above

6. Hirschi and Stark analyzed the association between church attendance and delinquency by controlling for gender. They did this to test for
   a. random selection.
   b. sampling bias.
   c. spuriousness.
   d. simple facts.
   e. population oversampling.

7. Recent *World Values Survey* data indicate that married persons in western nations are happier than unmarried persons. However, among non-western nations, there is little difference in the happiness of married and unmarried persons. This appears to be an example of a/an __________ effect.
   a. spurious
   b. random
   c. simple
   d. contextual
   e. independent

8. Network analysis
   a. is a research method.
   b. studies relationships among group members.
   c. attempts to describe group properties.
   d. uses networks as the unit of analysis.
   e. all of the above

9. Commune comes from the Latin word meaning
   a. communication.
   b. conflict.
   c. common.
   d. covetous.
   e. cooperation.

10. Zablocki initially believed that strong love networks among commune members would
   a. effect neither the group's stability nor its solidarity.
   b. increase the group's stability or solidarity.
   c. decrease the group's stability and solidarity.
   d. increase the group's stability but decrease solidarity.
   e. decrease the group's stability but increase solidarity.

11. In Zablocki's study on communes, group stability was the __________ variable.
  a. independent
  b. random
  c. spurious
  d. dependent
  e. contextual

12. The high membership turnover and disintegration rates observed in the communes studied by Zablocki were attributed to
  a. weak love networks.
  b. jealousy and competition.
  c. the equal sharing of financial resources.
  d. all of the above
  e. none of the above

13. The three major components of social systems are
  a. time order, correlation and spuriousness.
  b. samples, populations and censuses.
  c. separate parts, interdependence, stability and equilibrium.
  d. experimental research, survey research and network analysis.
  e. involution, status groups and class.

14. An inflationary economy may limit the resource opportunities a family may provide for its members. What property of systems does this illustrate?
  a. interdependence
  b. spuriousness
  c. random selection
  d. functional equivalence
  e. evolutionary progress

15. In many societies religions address questions of ultimate concern and provide comfort. Religion is an example of a/an
  a. spurious effect.
  b. closed system.
  c. social institution.
  d. independent variable.
  e. dependent variable.

16. Societies are not static; they fluctuate and change constantly. This means that societies tend to be __________ systems.
  a. open
  b. closed
  c. neutral
  d. positive
  e. negative

17. A system of parts, consequences, intent, purpose, equilibrium, disruption and alternative parts are concepts primarily associated with this macrosociological perspective.
   a. conflict theory
   b. social evolution
   c. symbolic interactionism
   d. exchange theory.
   e. functionalism.

18. A family that is comprised of several adult couples and their children is known as a/an
   a. extended family.
   b. single parent family.
   c. nuclear family.
   d. commune.
   e. household.

19. Since friends may exert as much influence on a teenager's life as will the family, functionalists may argue that friends and family are
   a. dysfunctions.
   b. status groups.
   c. symbols of progress.
   d. functional alternatives.
   e. proportional facts.

20. Adaptation and survival are two concepts primarily associated with the __________ perspective.
   a. conflict
   b. social evolutionary
   c. rational choice
   d. functionalist
   e. symbolic inteactionist

21. Social evolutionary theorists maintain that over time, societies have become
   a. smaller.
   b. more urban.
   c. more complex.
   d. a and b
   e. b and c

22. Many nineteenth century social evolutionists argued that change is
   a. random.
   b. subject to chance.
   c. inevitable and progressive.
   d. guaranteed and dysfunctional.
   e. the primary source of conflict among status groups.

23. Marx and Weber are generally associated with this sociological perspective
   a. social evolutionary theory.
   b. rational choice theory.
   c. exchange theory.
   d. functionalism.
   e. conflict theory.

24. Conflict theories are primarily concerned with
   a. the distribution of power within society.
   b. networks of love and jealousy.
   c. adaptation and survival.
   d. the evolution of progress.
   e. identifying functional alternatives.

25. What sociological perspective maintains that the norms of society are created and maintained by the ruling class?
   a. social evolutionary perspective.
   b. conflict perspective.
   c. exchange perspective.
   d. functional perspective.
   e. symbolic interactionist perspective.

26. According to Weber racial and ethnic groups and religious groups are examples of
   a. contextual effects.
   b. social classes.
   c. spuriousness.
   d. involution.
   e. status groups.

27. In comparative sociological research, the __________ is generally the unit of analysis.
   a. society
   b. city
   c. institution
   d. small group
   e. individual

28. Among non-industrial societies conflict, competition and violence tend to be higher in
   a. communal societies.
   b. progressive societies.
   c. alternative societies.
   d. factional societies.
   e. contextual societies.

29. Paige proposed that patterns of violence in non-industrial societies would be regulated by
    a. kinship ties and urbanization.
    b. residence patterns and technological progress.
    c. group size and age.
    d. gender and age.
    e. kinship ties and residence patterns.

30. The matrilocal rule of residence states that a married couple will
    a. live alone and establish a separate household.
    b. live near the wife's family.
    c. live near the husband's family.
    d. first live with the husband's family before finally moving in with the wife's family.
    e. first live with the wife's family before finally moving in with the husband's family.

True/False

Answers and page references are provided at the chapter end.

1. Group size is an attribute of social structure.

2. Census is to sample as whole is to part.

3. The three criteria of causation are census, sample and random selection.

4. Compared to experimental research design, the survey research method relies more heavily on data obtained from personal interviews and questionnaires.

5. Initially, Hirschi and Stark observed that the relationship between church attendance and delinquency is spurious. Later, Stark realized that the conflicting findings of other researchers were due to the presence of contextual effects.

6. When Zablocki applied network analysis to the analysis of intense love attachments in communes, he observed that intense love attachments promote group stability and group solidarity.

7. In many societies families address important social needs like emotional support and child rearing. Families are therefore examples of social institutions.

8. Social evolutionary theories focus on the intent and purpose of different social phenomena; whereas, functionalist theories focus on how social phenomena survive by adapting to change.

9. Marx associated conflict with competition among status groups while Weber associated conflict with social classes.

10. Paige's comparative study of non-industrial societies reveals that matrilocal societies tend to be factional and are characterized by high levels of internal violence.

Short Answer Questions

These short answer questions are provided to test your knowledge and understanding of the basic sociological concepts presented in Chapter 4. Page references for answers are included at the chapter end.

1. Distinguish simple and proportional facts by providing an example of each.

2. Identify several techniques researchers employ to improve a sample's accuracy.

3. How are spuriousness and contextual effects demonstrated in the study on church attendance and delinquency by Hirschi and Stark?

4. According to Zablocki what impact does strong love networks have on communal group solidarity? Were the observed findings the expected results?

5. Identify three important properties of social systems.

6. Stinchcombe maintains that functionalist theories include three major components. Identify these components.

7. According to the functionalist perspective, how might extended families provide social equilibrium in less developed societies?

8. What are some of the major criticisms of the social evolutionary perspective?

9. How do Marx and Weber differ in their approach to the study of conflict?

10. Distinguish the patrilocal and matrilocal rule of residence.

Essay Questions

These questions are designed to test your understanding of key sociological concepts presented in Chapter 4 and your ability to apply these insights to concrete situations.

1. A researcher wants to study student's perceptions of safety on a large university campus. The researcher wants to base the study on a sample. How would the researcher select a sample that is representative of the university population?

2. A study yields surprising findings. How would the researcher go about determining whether the findings are due to the presence of spurious relationships or contextual effects?

3. Select three social institutions and demonstrate how they are independent and promote stability and equilibrium within society.

4. How do the functionalist, social evolutionary and conflict perspectives differ in their understanding of the role of equilibrium and change in society?

5. What important insights could be gained from a comparative sociological study of key social institutions like the family, religion and education? What empirical data sources could be consulted to provide comparative information on these institutions?

Answers

Multiple Choice

1. d (97)
2. b (98)
3. e (98-99)
4. a (99)
5. d (100)
6. c (100-101)
7. d (102-103)
8. e (104)
9. c (104)
10. b (105-106)
11. d (106)
12. b (106-107)
13. c (108-109)
14. a (108-109)
15. c (108)
16. a (109)
17. e (110-111)
18. a (110)
19. d (110-111)
20. b (111-112)
21. e (111)
22. c (112)
23. e (112-113)
24. a (112-114)
25. b (112)
26. e (113)
27. a (114)
28. d (117)
29. e (117)
30. b (118-119)

True/False

1. T (97)
2. T (98-99)
3. F (100-101)
4. T (100)
5. T (101-103)
6. F (106-107)
7. T (108)
8. F (110-112)
9. F (112-113)
10. F (118-119)

Short Answer

1. (96-98)
2. (99)
3. (101-103)
4. (104-107)
5. (108-109)
6. (110)
7. (110)
8. (112)
9. (112-113)
10. (118)

CHAPTER 5

BIOLOGY, CULTURE, AND SOCIETY

Extended Abstract

Researchers have debated the impact of heredity and environment on human behavior continuously. Are we a product of our genetic background? Do environmental factors shape and constrain behavioral choices? Rather than casting the nature-nurture debate within a nature versus nurture context, Stark reviews recent research that demonstrates the dynamic interplay of nature and nurture. This interplay is addressed in twin studies, the emerging field of behavioral genetics, improvements in quality of life and standard of living, comparative primate studies, tool-making among non-human species and language acquisition among primates.

Chromosomes and genes, chains of genetic codes, specify an organism's physical traits. However, human genetic potential, the genotype, may be restricted by environmental factors such as quality of life, diet, sanitation and health care. An organism's actual genetic outcome is its phenotype. The phenotype is a result of the dynamic interplay of an organism's genotype and its environment.

Behavioral geneticists maintain that some human behavioral characteristics, like mental illness and alcoholism, are influenced by genetic inheritance. Studies of identical twins provide insight in this area. However, to what degree can the behavioral outcomes of identical twins can be attributed to the same genetic composition, and how much can be attributed to similar environmental contexts? Research involving adopted twins suggests that environmental factors modify human potential. Also, improvements in a society's quality of life and standard of living may alter human genetic expression. Sanitation, diet and better health care impact life expectancy, infant mortality and provide a context to extend human genetic potential. Stark notes that discrimination among groups is reduced when environments, which allow for the objective expression and measurement of human performance, are maintained. Sports and entertainment are offered as examples of performance based environments.

The dynamic interaction among biology, culture and society is illustrated by recent studies assessing the relationship among hormones, DNA, behavior and culture. The massive "Vietnam Veterans Health Status Study" conducted by the Centers for Disease Control indicate that strong associations exist among high testosterone levels and the risk of divorce, abusive behavior, sexual promiscuity and alcoholism. These associations remain when control variables are introduced. Likewise, DNA studies suggest that throughout history women have been more geographically mobile. More recently, the Jewish kinship ties of the Lemba of southern Africa have been confirmed through DNA analysis.

The final issue addressed in this chapter is the distinctiveness of human behavior. Social scientists maintain that the uniqueness of human behavior is rooted in the ability to create, modify and transmit culture. However, studies of primate behavior by Goodall, the Harlows and the Gardiners challenge this assumption. Based on extensive field observation of primate behavior, Goodall demonstrates that chimpanzees are toolmakers and that they will kill other monkeys. Goodall further notes that adult chimps teach younger chimps how to make tools. Thus, chimpanzees create, modify and transform culture. Recent studies also indicate that female crows can make tools. Research by the Harlows indicates that interaction with other primates is critical to the emotional development of primates. Monkeys reared in isolation were unable to participate fully in "normal" monkey social environments. Additional research by the Harlows indicates that cuddling rather than nursing monitors the primary mother-offspring tie among primates. Finally, studies by the Gardiners provide evidence that language acquisition is not a unique human trait. Chimpanzees are able to master American Sign Language and can teach signing to adopted offspring.

Key Learning Objectives

After a thorough reading of Chapter 5, you should be able to:

1. better comprehend the complex interaction among biology, culture and society.
2. state the association among chromosomes, genes and heredity.
3. clarify the interaction between genotype and phenotype.
4. identify the major research interests of behavioral geneticists.
5. demonstrate why identical twin research may provide key insights into the association among heredity, environment and behavior.
6. provide examples of how environmental factors like diet, sanitation and health care can suppress genetic potential.
7. state how individual and group opportunity and performance may be related to reductions in discrimination.
8. assess the impact of hormones on human behavior by reviewing findings from studies on behavior and variations in testosterone levels.
9. indicate how DNA research may be used to verify kinship ties among geographically diverse cultural groups.
10. critically evaluate the distinctiveness of human behavior.
11. state how Leakey's studies of fossil remains provide a clearer understanding of human origins.
12. identify the major contributions of Goodall to primate studies and field observation methods.
13. demonstrate why interaction with other primates is critical to primate development.
14. state why tool-making is not a distinctively human trait.

15. evaluate studies on primate language acquisition.

Chapter Outline

I. Nature Versus Nurture or Nature and Nurture
   A. Instincts and Behavior: McDougall's Social Psychological Approach
   B. Environment and Behavior: Margaret Mead
   C. The Interaction of Nature and Nurture

II. Heredity and Behavioral Genetics
   A. The Genetic Code: Chromosomes and Genes
   B. Genetic Potential and Performance: Genotype and Phenotype
   C. Behavioral Genetics: Human Behavior and Inherited Traits
      1. Studies involving IQ and personal traits among identical twins.
      2. Studies involving identical twins and adoptive parents.
      3. Evaluating the impact of shared inheritance and shared environment.

III. The Industrial Revolution and Changes in the Quality of Life
   A. Increases in Human Physical Growth: Genes and Environment.
   B. Environmental Suppression of Human Genetic Potential
      1. Infant mortality and changes in quality of life.
      2. Life expectancy and changes in quality of life.
   C. Opportunity, Individual and Group Performance and Discrimination
      1. Minority group experiences in sports.
      2. Minority group experiences in entertainment.

IV. Genetics, Environment and Behavior: Case Studies
   A. Evaluating the Impact of Hormones on Human Behavior: Testosterone Studies
   B. The "Vietnam Veterans Health Status Study"
      1. Testosterone levels linked with certain behavioral risks.
      2. Testosterone effects remain when statistical controls are introduced.
   C. DNA Studies and Hereditary Links
      1. Women have been more geographically mobile than men.
      2. Jewish kinship ties among the Lemba of southern Africa established.

V. Primate Studies and Human Distinctiveness
   A. Traditional Social Science Assumptions Concerning Human Behavior
      1. Humans are an intellectually superior species.
      2. Humans are a morally and ethically inferior species.
   B. Jane Goodall: A Pioneer in Primate Studies
      1. Toolmaking observations reveal that primates are capable of creating and transmitting culture.
      2. Primates kill other primates.
      3. Developed principles of field observation based on meticulous record keeping and critical evaluation of details.

C. Tool-making among Female Crows

D. Isolation and Primate Behavioral Development: Harlow Studies
   1. Early interaction with other primates has a major impact on behavioral development.
   2. The primary mother-offspring tie is based on cuddling rather than nursing.

E. Studies on Primate Language Acquisition: The Gardiners
   1. Primates communicate primarily through gestures and body movements.
   2. Primates can learn and use American Sign Language.
   3. Primates can teach sign language to their young.

Key Terms

Based on your reading of Chapter 5, you should be able to define and illustrate the key sociological concepts listed below. Page numbers are provided in parentheses as reference points.

Instinct (125)
Chromosomes (127)
Genes (127)
Genotype (128)
Phenotype (128)
Behavioral Genetics (129)
Identical (Monozygotic) Twins (131)
Fraternal (Dizygotic) Twins (132)
Hormones (135)
Testosterone (135)
Estrogen (135)
DNA (136)
Tool (139)

Key Research Studies

Listed below are key research studies cited in Chapter 5. Familiarize yourself with the major findings of these studies. Page references are provided in parentheses.

McDougall: developed social psychological approach that emphasized the instinctive basis of human behavior (125).

Blalock: identified links among opportunity structures like sports, individual and group performance and discrimination (131).

Schuckit: studied association among alcoholism of biological and adoptive parents and alcoholism of adopted children. The environmental influence provided by the adoptive parent is identified as a stronger factor impacting alcoholism among adopted children (132).

Vietnam Veterans Study: testosterone levels are linked to a higher risk of being divorced, sexual promiscuity, alcohol and drug abuse and other behavioral outcomes (136).

Lemba Origins: DNA analysis provides key link in establishing Jewish kinship ties of a cultural group presently residing in southern Africa (137).

Goodall: conducted extensive field observation of primates. Demonstrated that chimpanzees are able to create and transmit culture. Chimpanzees create tools and teach their offspring to make and use them (138-140).

The Harlows: monkeys reared in isolation are unable to interact effectively with other primates. The primary mother-child tie is associated with cuddling rather than nursing (141-142).

The Gardiners: primates communicate through the use of gestures and are able to learn and use American Sign Language. Primates can teach sign language to their young (143-144).

Info Trac Search Words

Enter these search terms to conduct more extensive investigations of key topics introduced in Chapter 5.

Nature and Nurture

Select one of the studies listed. Identify the author's perspective. Does the author view behavior as primarily a product of nature, nurture or the interaction of nature and nurture?

Testosterone

Subdivisions: Research

Look at one of the studies listed. What behavioral characteristics are associated with higher testosterone levels?

DNA

Subdivisions: Usage

Louis Leakey

Tools and Primates

Multiple Choice

Answers and page references are provided at the chapter end.

1. In *An Introduction to Social Psychology*, McDougall argued that behavior is a product of
   a. environment.
   b. culture.
   c. economics.
   d. religion.
   e. instincts.

2. Research has demonstrated that human infants
   a. are born with the ability to imitate facial expressions.
   b. learn how to imitate facial expressions when they are three months old.
   c. learn how to imitate facial expressions when they are six months old.
   d. learn how to imitate facial expressions when they are eight to ten months old.
   e. do not learn to imitate facial expressions until they are a year old.

3. Stark argues that the nature-nurture debate is best understood when researchers note
   a. that environmental influences on behavior are stronger than heredity .
   b. that hereditary influences on behavior are stronger than environmental factors.
   c. behavior is not influenced by heredity or environment.
   d. that behavior is influenced by hereditary and environmental factors.
   e. none of the above

4. In humans twenty-three chromosomes from the mother combine with __________ chromosomes from the father.
   a. one
   b. two
   c. twenty-three
   d. fifty
   e. over one hundred

5. Chromosomes are comprised of
   a. genes.
   b. DNA.
   c. culture.
   d. a and b
   e. b and c

6. Genotype addresses __________; whereas, phenotype is an expression of __________.
   a. genetic potential; genetic performance
   b. genetic performance; genetic potential
   c. environmental potential; environmental performance
   d. environmental performance; environmental potential
   e. all of the above

7. Behavioral geneticists maintain that __________ is influenced by inheritance.
   a. intelligence
   b. mental illness
   c. alcoholism
   d. aggressive behavior
   e. all of the above

8. When behavioral geneticists try to demonstrate that a trait is inherited, they need to demonstrate that the trait is
   a. more common among blood related rather than unrelated individuals.
   b. equally common among blood related and unrelated individuals.
   c. more common among unrelated individuals than blood related individuals.
   d. more common among males than females.
   e. more common among females than males.

9. According to Stark minority groups in North America have tended to experience success first in the fields of
   a. politics and the military.
   b. sports and entertainment.
   c. religion and sports.
   d. politics and entertainment.
   e. sports and politics.

10. Blalock maintains that minority group discrimination is
   a. lower in occupations where individual performance is easily measured.
   b. higher in occupations where individual performance is easily measured.
   c. not influenced by individual performance environments.
   d. difficult to document and prove.
   e. no longer a major issue in U.S. society.

11. Monozygotic twins are
   a. identical twins produced when two ova are impregnated.
   b. fraternal twins produced when two ova are impregnated.
   c. identical twins produced when an impregnated ova splits.
   d. fraternal twins produced when an impregnated ova splits.
   e. dizygotic twins produced when ova are artificially inseminated.

12. Fraternal twins
   a. may not be of the same sex.
   b. share the same genetic inheritance.
   c. are genetically dissimilar.
   d. a and b
   e. a and c

13. Although the IQs of identical twins are not the same, they may be very similar. This may be due to
   a. the sharing of the same genetic structures.
   b. being reared in similar environments.
   c. birth order effects.
   d. a and b
   e. none of the above

14. Which one of the following statements is false?
    a. The IQs of identical twins reared apart are very similar.
    b. Only 30 % of the variation in IQs of twins reared apart is due to genetic variation.
    c. Children of a schizophrenic parent are more likely to develop schizophrenia.
    d. Adoptive children of schizophrenics are highly unlikely to become schizophrenic.
    e. Seventy percent of the variation in intelligence among twins reared apart is due to environmental factors.

15. Compared to their parents and grandparents, adult children in industrial societies are
    a. taller.
    b. shorter.
    c. the same height.
    d. more likely to have a shorter life expectancy.
    e. more likely to die from infectious disease.

16. Throughout the twentieth century U.S. males
    a. have grown taller while females have become shorter.
    b. have become shorter while females have grown taller.
    c. and females have remained the same height.
    d. and females have become shorter.
    e. and females have grown taller.

17. Declines in infant mortality in industrial societies has been attributed to
    a. better sanitation.
    b. better diet.
    c. better health care.
    d. all of the above
    e. none of the above

18. Testosterone and estrogen are primarily examples of
    a. environmental suppressors.
    b. instincts.
    c. hormones.
    d. DNA.
    e. chromosomes.

19. Many early studies assessing the impact of testosterone levels on human behavior were based on
    a. small samples.
    b. large samples.
    c. complete populations.
    d. telephone interviews.
    e. field observations only.

20. According to the Vietnam Veterans Study, men with higher testosterone levels are
    a. less likely to get divorced.
    b. less likely to be faithful to their wives.
    c. more likely to get married.
    d. more likely to obtain high-status employment.
    e. more likely to have problems with alcohol and drug abuse.

21. DNA is a __________ concept.
    a. economic
    b. sociological
    c. cultural
    d. biological
    e. political

22. Recent DNA studies confirm that
    a. Asians have been more geographically mobile than Europeans.
    b. Christians have been more geographically mobile than Jews.
    c. women have been more geographically mobile than men.
    d. minority groups have been more geographically mobile than dominant groups.
    e. people in agricultural societies have been more geographically mobile than people in industrial societies.

23. DNA analysis was employed recently to demonstrate that the Lemba of southern Africa are descended from
    a. Europeans.
    b. Jews.
    c. Asians.
    d. Australians.
    e. Indonesians.

24. The archeological discoveries of this researcher indicate that human ancestors walked the face of the earth millions of years ago
    a. Jean-Jacques Rousseau.
    b. Margaret Mead.
    c. Louis Leakey.
    d. Harry Harlow.
    e. Jane Goodall.

25. Social scientists have traditionally maintained that humans
    a. are an intellectually superior species.
    b. can create culture.
    c. can transmit culture to their offspring.
    d. are a morally and ethically inferior species.
    e. all of the above

26. Lorenz argued that murder is a uniquely human characteristic. This behavior was associated with the belief that
    a. humans started to live in cities in the early stages of the evolutionary process.
    b. women are stronger and genetically superior to men.
    c. men are generally taller than women.
    d. humans became meat eaters in the later stages of the evolutionary process.
    e. humans were able to create a legal code.

27. Goodall discovered that primates
    a. make tools.
    b. are intellectually superior to humans.
    c. kill other primates.
    d. a and b
    e. a and c

28. The primate research conducted by the Harlows suggests that primate infants reared in different types of isolated environments
    a. live longer.
    b. experienced normal social development.
    c. are less aggressive.
    d. grow to be taller and heavier.
    e. experienced problems interacting with other primates.

29. Primates generally communicate through the use of
    a. gestures.
    b. vocal speech.
    c. tools.
    d. smell.
    e. isolation and avoidance.

30. The studies by the Gardiners, Fouts and others indicate that
    a. primates can learn sign language.
    b. primates can teach their young to sign.
    c. primates can communicate through written language.
    d. a and b
    e. a and c

True/False

Answers and page references are provided at the chapter end.

1. Margaret Mead's research indicated that human behavior is primarily instinctive.

2. Genotype is to phenotype as genetic potential is to genetic performance.

3. Behavioral genetics research places more emphasis on studies of fraternal twins.

4. Improved sanitation and improved access to health care has been associated with higher rates of infant mortality in industrial societies.

5. Testosterone and estrogen are sex hormones that are believed to influence human behavior.

6. One of the major findings of the Vietnam Veterans Health Status Study is that high testosterone levels are associated with a higher risk of being unemployed.

7. DNA research suggests that throughout history men have been more geographically mobile than women.

8. Goodall's research on primate tool-making indicates that tool-making is an example of instinctive behavior.

9. The Harlow primate studies suggest that the mother-child bond is based on cuddling rather than nursing.

10. Language acquisition is not a unique, distinctive human characteristic.

Short Answer Questions

These short answer questions are provided to test your knowledge of the basic sociological concepts presented in Chapter 5. Page references for answers are included at the chapter end.

1. What was the nature of the association between heredity and behavior according to McDougall?

2. Distinguish genotype and phenotype.

3. Identify several examples of behavior that behavioral geneticists tend to associate with inheritance.

4. Why are twin studies an important feature of behavioral genetics research?

5. According to Blalock what is the nature of the association among opportunity, individual and group performance and discrimination?

6. Identify several changes brought about by the Industrial Revolution that would tend to lower infant mortality and increase life expectancy.

7. What types of human behaviors have been associated with high levels of testosterone production?

8. Identify three rules of field observation developed by Goodall in her primate studies.

9. Why is tool-making not a distinctively human trait?

10. What is the primary mode of communication utilized by primates?

Essay Questions

These questions are designed to test your understanding of key sociological concepts presented in Chapter 5 and your ability to apply these insights to concrete situations.

1. Often the nature-nurture debate is cast within a nature versus nurture context. Make a case for casting the debate within a nature and nurture framework.

2. How are contemporary behavioral genetics studies similar and different from late nineteenth and early twentieth century social psychological studies on instincts and behavior?

3. Demonstrate how findings from twin studies reflect the dynamic interdependence of heredity, environment and behavior?

4. How could DNA analysis aid researchers to gain a better understanding of the historical development of global migration patterns?

5. Are humans the only species capable of creating, modifying and transmitting culture? Provide examples to support your position.

Answers

Multiple Choice

1. e (125)
2. a (127)
3. d (127)
4. c (127)
5. d (127)
6. a (128)
7. e (129)
8. a (129)
9. b (130)
10. a (131)
11. c (131-132)
12. e (132)
13. d (131-132)
14. b (132)
15. a (133-134)
16. e (133-134)
17. d (134-135)
18. c (135)
19. a (135-136)
20. e (136)
21. d (136)
22. c (136-137)
23. b (137)
24. c (138-139)
25. e (138)
26. d (138)
27. e (139-140)
28. e (141-142)
29. a (143)
30. d (143-144)

True/False

1. F (126)
2. T (128)
3. F (129-132)
4. F (134-135)
5. T (135)
6. T (136)
7. F (136-137)
8. F (139-140)
9. T (141-142)
10. T (143-144)

Short Answer

1. (125)
2. (128)
3. (129)
4. (129-132)
5. (131)
6. (134-135)
7. (135-136)
8. (140)
9. (139-141)
10. (143-144)

CHAPTER 6

SOCIALIZATION AND SOCIAL ROLES

Extended Abstract

Socialization is the process of becoming a group member, a social being. This process occurs throughout life and involves learning group norms and assuming behavioral roles. The nature and the extent of the contact with others are critical to successful human development and maturation. Studies of children and primates reared in isolation indicate that early social interaction stimulates normal development. However, accelerating technologies, such as the "Mozart Effect," have questionable impact on human development.

Since socialization occurs throughout life, how do humans learn? Stimulus response (S – R) theories indicate that humans learn passively by responding to environmental stimuli and repeating responses that are reinforced. On the other hand, Piaget argued that humans develop reasoning skills at different ages, which enable them to play a more active role in the learning process. At the earliest stage of cognitive development (birth – 2 years), children learn the rule of object permanence. At later ages they learn to empathize with others (overcome egocentrism), develop empirical reasoning skills (rule of conservation) and eventually comprehend abstract concepts (ages 12 and over). Everyone does not necessarily attain this last stage.

Piaget also proposed that the different stages of cognitive development are tied to different stages of brain maturation. Recent research on language acquisition and the language instinct sheds light on this hypothesized association. Rather than merely storing words and sentences and imitating words previously heard, Chomsky maintains that children possess an inborn Universal Grammar. Young children learn languages quickly and are able to produce grammatically correct expressions that have not been heard previously. The creation of new languages (creoles), based on a jargon of made-up words (pidgin), provides additional support for an inborn language instinct. This is illustrated in Kegl's study on the creation of a specialized sign language among Nicaraguan deaf children. Rather than learning to lip-read or speak Spanish, these deaf children communicate with each other by creating their own system of gestures and pantomimes.

As humans mature cognitively, they develop a sense of self and personality. This sense of self emerges as individuals interact with others, learn how to place themselves in another person's place and can judge how others are responding to the way they act. In fact, Bain observed that children tend to develop other-related words before developing self-related words. Personality, the distinctive way a person thinks, feels and acts, is a product of extensive social interaction. Some aspects of personality appear to be more universal while others vary substantially among different cultural settings. Some aspects of personality are unique since no two people are exactly alike.

Given the tremendous variation in human personality, social scientists place great emphasis on the interplay of culture, social structure and socialization. Writing from a cultural determinism perspective, Boas maintained that culture can shape and modify personality into almost any form. Margaret Mead, a student of Boas, attempted to illustrate this point in her classic study of sex roles between the Arapesh and Mundugumor of New Guinea. She observed that Arapesh men and women were reared to exhibit "feminine" temperaments, like gentleness;

whereas, Mundugumor men and women were reared to exhibit "masculine" temperaments like aggressiveness. Mead concluded that sex roles are culturally based rather than biologically based and vary from culture to culture.

Offering a critique of Mead's findings and the cultural determinism perspective, Stark cites data from the *Standard Cross-Cultural Sample* to demonstrate that masculine temperaments are encouraged in the socialization of boys and girls when societies exist in close proximity to each other. Upon closer examination, one learns that the Mundugumor lived in close proximity to other people while the Arapesh were more isolated. Thus, the Mundugumor may find it more advantageous to teach their members to protect their assets. This suggests that environmental as well as cultural factors shape personality development. Also, Stark notes that socialization continues throughout life. The impact of the socialization process on personality development extends well beyond early child rearing practices.

The remainder of Chapter 6 is devoted to a discussion of differential socialization. Within any given society, all members are not encouraged to assume the same social roles. Differences exist even in the same family where daughters and sons and older and younger children are reared differently. Two examples of differential socialization involve parental child rearing style preferences and sex-role socialization.

Kohn has conducted extensive research on the impact of parental work environment on child rearing style. In the early stages of his research, Kohn noted that working-class parents encouraged their children to please others and conform to the expectations of others; whereas, middle-class parents stressed independence. Upon closer examination, Kohn noted that conformity was stressed more among parents with jobs in highly supervised work environments while parents who worked in settings that allowed more creativity stressed independence. The difference in child rearing style preferences reflects differences in parents' occupational experiences rather than social class effects. In later follow-up research, Kohn observed that persons with more flexible personalities seek out jobs that require less supervision. Kohn's findings have been replicated cross-nationally. The strong impact of parental work experience on child rearing practices is consistent.

Before moving to sex-role socialization, Stark provides a brief excursus on Goffman's studies on role performance. Role performance addresses how people actually portray the role they are playing. Persons care about their image and will manipulate their environment (impression management) to convey a favorable image. Conversely, people ignore behavioral miscues (studied nonobservance) that could be potentially embarrassing.

Sex-role expectations in many societies are very distinct. Data from the *Standard Cross-Cultural Sample* indicate that women have less political power, that boys are more often socialized to be tough and aggressive and that women are given more authority in the care of younger children. Men are given more authority with older children. But how are these roles learned and reinforced? In several studies involving the identification of gender-neutral characters in children's stories, DeLoach, Cassidy and Carpenter note that most mothers use masculine labels to identify the characters. The bias is still present when the gender-neutral characters are portrayed in sex-typed activities. These researchers concluded that children's stories should portray female characters that challenge the traditional sex-role stereotypes. Finally, observing children's play, Richer notes that younger boys and girls (aged 3-4) play games together easily while older children (aged 6-14) do not. This suggests that the gender preferences of older children are learned.

Key Learning Objectives

After a thorough reading of Chapter 6, you should be able to:

1. describe the key features of the socialization process and indicate how the socialization process continues throughout life.
2. indicate how environmental factors can suppress human development.
3. specify how learning takes place within a stimulus-response (S – R) context.
4. identify the cognitive crises associated with each of Piaget's four stages of cognitive development.
5. critically evaluate studies that suggest that language acquisition is instinctive.
6. demonstrate how people create a new language (creole) through the development of a pidgin.
7. appreciate the universal and unique nature of human personality.
8. critically evaluate the claims of the cultural determinism tradition of Franz Boas and Margaret Mead.
9. provide examples of differential socialization.
10. distinguish the child rearing practices of parents experiencing different work environments.
11. identify the type of information that can be gained from longitudinal studies.
12. note cross-national differences in child rearing values.
13. describe how individuals manipulate their environment to enhance role performance.
14. identify the similarities and differences in gender role expectations in non-industrial and industrial societies.
15. evaluate the roles parents and peers play in gender socialization.

Chapter Outline

I. The Socialization Process and Human Development
    A. Socialization: The Process of Becoming a Group Member
    B. Isolation and Human Development

1. Feral children and the nature of human contact.
2. Primate isolation and primate development (Harlow studies).
3. Institutional care and isolation (Skeels and Dye).

C. Accelerated Development and the Proposed "Mozart Effect"
1. Unsubstantiated claims concerning impact of classical music on human development.
2. Physiological constraints on human development.

II. Theories of Cognitive Development

A. Stimulus Response (S – R) Theories: A Passive Approach
1. Learning through responses to external stimuli.
2. Shaping behavior through the use of reinforcements.

B. Piaget's Cognitive Stages: An Active Approach
1. Acquiring basic reasoning skills.
2. Learning that objects continue to exist when they are "out of sight."
3. Overcoming egocentrism by seeing things from another's perspective.
4. Developing empirical and abstract reasoning skills.
5. Cognitive development and brain maturation, a key interaction.

III. Language Acquisition and the Language Instinct

A. Chomsky's Universal Grammar
1. Persons possess an instinctive awareness of grammatical rules.
2. Children acquire different grammars quickly.

B. Linguistic Capacity and the Creation of Pidgins and Creoles

C. Nicaraguan Sign Language: A New Creole

IV. The Self, Personality and Culture

A. Discovering the Self by Taking on the Role of the Other: Bain and Flavell

B. Personality: Universal and Distinctive Patterns of Thought, Feeling and Action

C. Cultural Determinism: The Work of Franz Boas and Margaret Mead
1. Culture shapes personality primarily through child rearing practices.
2. Sex roles are culturally determined and vary from culture to culture.
3. Stark's critical assessment.
   a. Child rearing practices influenced by proximity of other cultural groups.
   b. Socialization continues throughout life.

V. Differential Socialization and Societal Expectations

A. Birth Order Expectations: Differential Socialization within Families

B. Parental Occupational Experiences and Child Rearing Expectations: Kohn
1. Parents in highly supervised work environments encourage children to conform to the expectations of others.
2. Parents in more casual work environments stress self-expression and independence.
3. Longitudinal studies indicate that persons with flexible, self-directed personalities seek jobs with less supervision.

4. Cross-national studies link parents' work experiences and child rearing practices.

C. An Excursus on Role Performance: Goffman

1. People manipulate social environments to present favorable images (impression management).
2. People tend to ignore behavioral miscues (studied nonobservance).

D. Sex-Role Socialization Patterns

1. Subordination of women in non-industrial societies.
2. Sex-role expectations in industrial societies.
3. Gender-neutral children's story characters, sex-typed activities and male identification bias (DeLoach, Cassidy and Carpenter).
4. Sex segregated play among older children and learned behavior (Richer).

Key Terms

Based on your reading of Chapter 6, you should be able to define and illustrate the key sociological concepts listed below. Page numbers are provided in parentheses as reference points.

Feral Children (149)
Socialization (150)
Mozart Effect (152)
Stimulus Response (S-R) Theory (153)
Cognitive Structures (153)
Sensorimotor Stage (154)
Rule of Object Permanence (154)
Preoperational Stage (154)
Egocentrism (154)
Concrete Operational Stage (154)
Role of Conservation (154)
Formal Operational Stage (155)
Language Instinct (155)
Pidgin (157)
Creole (157)
Self (158)
Personality (159)
Cultural Determinism (160)
Ethnographic Atlas (162)
Standard Cross-Cultural Sample (162)
Differential Socialization (163)
Adult Socialization (166)
Longitudinal Study (166)
Role Performance (168)
Impression Management (168)
Stage (168)
Backstage (168)
Teamwork (168)
Studied Nonobservance (168)
Deviant Role (169)
Sex-Role Socialization (169)

Key Research Studies

Listed below are key research studies cited in Chapter 6. Familiarize yourself with the major findings of these studies. Page references are provided in parentheses.

Skeels and Dye: interaction with others stimulates human development within an institutionalized setting (151-152).

Rauscher, Shaw and Ky: hypothesized that listening to classical music, the "Mozart

Effect," would improve IQ. Researchers have been unable to replicate findings (152-153).

Piaget: developed a theory of cognitive stages grounded in the comprehension of basic rules of reasoning such as object permanence and empirical judgment (153-155).

Brown and Bellugi: children experiment with various speech patterns as they attempt to discover a language's basic grammatical rules (156).

Chomsky: children possess an inborn universal grammar that enables them to learn and use languages easily at young ages (156).

Kegl: the creation of Nicaraguan Sign Language by deaf children is a contemporary example of a pidgin and a creole (157-158).

Bain: children utilize other-related words before they acquire self-related words (159).

Flavell: demonstrated how older children (aged 14) differ from younger children (aged 8) in their ability to overcome egocentrism (159).

Boas and Mead: believed that culture is the major determinant of personality. Childhood socialization practices were viewed as a very strong determinant of personality (160-162).

Ethnographic Atlas and *Standard Cross-Cultural Sample*: data sources for studying behavioral trends in non-industrial societies (162, 170)

Kohn: parental work environment experiences influence child rearing strategies (165-167)

World Values Survey: data source for studying cross-national behavioral trends primarily among industrial societies (167, 171).

Goffman: illustrates how persons manipulate their environment (impression management and studied nonobservance) to enhance role performance (167-169).

DeLoach, Cassidy and Carpenter: identified a bias toward male sex-role identification of gender-neutral characters in children's stories (170-172).

Richer: analyzed play patterns of younger and older children and noted that the gender segregated play preferences of older children appear to be examples of learned behavior (172-173).

Info Trac Search Words

Enter these search terms to conduct more extensive investigations of key topics introduced in Chapter 6.

Music, Influence of
    Subdivisions: Social Aspects

Language Acquisition
    Subdivisions: Research

Margaret Mead
    Margaret Mead is one of the best-known cultural anthropologists. Select one of the references listed and identify some of her contributions to the general study of culture or the interaction of culture and personality.

Birth Order Effects

Gender Socialization
    Select one of the studies cited. What does your study say about the nature of gender socialization?

Multiple Choice

Answers and page references are provided at the chapter end.

1. Feral children are children reared
    a. in male single parent households.
    b. by grandparents.
    c. in isolation.
    d. in female single parent households.
    e. by persons other than blood relatives.

2. Socialization occurs during
    a. the early stages of infancy.
    b. early childhood.
    c. adolescence.
    d. adulthood.
    e. all the above

3. Skeels and Dye observed that children in an orphanage who had received personal attention from an older mildly retarded girl
    a. showed improvement in their IQ scores four years later.
    b. saw a decline in their IQ scores four years later.
    c. showed no change in their IQ scores four years later.
    d. were less able to adapt to change.
    e. more likely to prefer being isolated from other children.

4. Studies attempting to replicate the "Mozart Effect" have
   a. shown that listening to classical music significantly raises a young child's IQ.
   b. shown that listening to classical music significantly lowers a young child's IQ.
   c. demonstrated that the effect is stronger on males.
   d. demonstrated that the effect is stronger on females.
   e. been unsuccessful.

5. Stimulus response theory claims that behavior is
   a. a stimulus to external responses.
   b. a response to external stimuli.
   c. learned passively as the mind stores and retrieves information.
   d. a and b
   e. b and c

6. Each of the following is one of Piaget's stages of cognitive development except the
   a. formal operational stage.
   b. preoperational stage.
   c. concrete operational stage.
   d. differential socialization stage.
   e. sensorimotor stage.

7. Overcoming egocentrism is associated with the __________ stage.
   a. differential socialization
   b. preoperational
   c. formal operational
   d. sensorimotor
   e. concrete operational

8. A child is presented with two glasses of liquid. Each glass contains 4 ounces of liquid. One glass is an 8 ounce glass while the other glass is a 4 ounce glass. The child selects the 4 ounce glass because she thinks it contains more. Piaget would argue that the child
   a. has not learned the rule of object permanence.
   b. has overcome egocentrism.
   c. not learned the law of conservation.
   d. is utilizing abstract reasoning skills.
   e. is experiencing differential socialization.

9. Researchers studying cognitive development have concluded that approximately __________ of adults cannot comprehend abstract concepts and thus rely on literal interpretations of events.
   a. 5%
   b. 10%
   c. 25%
   d. 50%
   e. 67%

10. Stimulus response theories tie language acquisition to
   a. the trial and error search for grammatical rules.
   b. the language instinct.
   c. the imitation of things heard and said.
   d. all the above
   e. none of the above

11. In developing his concept of Universal Grammar, Chomsky maintains that
   a. children possess mental grammars, which enable them to combine words creatively.
   b. children develop complex grammars quickly and at a young age.
   c. children learn one language as easily as another.
   d. all of the above
   e. none of the above

12. A jargon of made-up words is known as a/an
   a. pidgin.
   b. language performance.
   c. feral grammar.
   d. universal grammar.
   e. creole.

13. The development of the Nicaraguan Sign Language illustrates how
   a. a creole is transformed into a pidgin.
   b. a pidgin is transformed into a creole.
   c. isolation stimulates the development of language skills.
   d. social interaction suppresses the development of language skills.
   e. young children are able to master the rule of object permanence.

14. The universal and distinctive pattern of thought, feeling and action is known as
   a. egocentrism.
   b. the sense of self.
   c. personality.
   d. feral children.
   e. socialization.

15. Boas and Mead maintained that personality is primarily shaped by
   a. history.
   b. economics.
   c. instincts.
   d. the sense of self.
   e. culture.

16. Boas argued that adult personality is effected most by
  a. instincts.
  b. early childhood socialization.
  c. the adolescent identity crisis.
  d. being reared in isolation.
  e. environmental factors such as urbanization.

17. Which of the following statements about Margaret Mead's ideas and research findings is true?
  a. Most of her studies were of African cultural groups.
  b. She maintained that gender roles are biologically determined.
  c. Adult socialization is more important than early socialization.
  d. Arapesh men exhibited feminine temperaments while Arapesh women exhibited masculine temperaments.
  e. Mundugumor men and women displayed masculine temperaments.

18. Data from the *Standard Cross-Cultural Sample* indicate that societies that live close to each other are
  a. less likely to engage in external war.
  b. more likely to prefer female infants.
  c. more likely to stress inflicting violence on outsiders.
  d. emphasize gentleness and cooperation in the socialization of males.
  e. emphasize gentleness and cooperation in the socialization of females.

19. Two major sources of anthropological data on non-industrial societies are the
  a. *Ethnographic Atlas* and the *Standard Cross-Cultural Sample.*
  b. the *Census* and the *Ethnographic Atlas*.
  c. *General Social Survey* and the *World Values Survey*.
  d. *Standard Cross-Cultural Sample* and the *General Social Survey*.
  e. *World Values Survey* and the *Census*..

20. Gender socialization is an example of
  a. feral socialization.
  b. differential socialization.
  c. the sex-role instinct.
  d. biological determinism.
  e. studied nonobservance.

21. Kohn initially argued that __________ parents encourage their children to please others by conforming to the expectations of others.
  a. underclass
  b. lower-upper class
  c. lower-class
  d. middle-class
  e. upper-upper class

22. In later research, Kohn discovered that the main factor influencing a parent's choice of child rearing strategies is
   a. age of the parent.
   b. gender of the child.
   c. social class.
   d. parent's work environment.
   e. mother's religious orientation.

23. A researcher conducted a study of alcohol use among students on a college campus. Twenty years later many of the former students who participated in the study were contacted to participate in a follow-up study. This researcher is trying to conduct a
   a. longitudinal study.
   b. cross-national study.
   c. study in isolation.
   d. latitudinal study.
   e. vertical panel study.

24. Kohn also observed that as time passes people working in highly structured environments become
   a. more flexible and more self-directed.
   b. less flexible and less self-directed.
   c. more flexible but less self-directed.
   d. less flexible but more self-directed.
   e. more dissatisfied and think about changing jobs.

25. Persons manipulate their environments in order to promote themselves in the most favorable manner possible. This is known as
   a. backstage role.
   b. studied nonobservance.
   c. impression management.
   d. cultural determinism.
   e. rule of status permanence.

26. A person bends over and splits their pants. Several people see it, but everyone ignores it. Goffman labels this
   a. creole.
   b. pidgin.
   c. costume management.
   d. prop management.
   e. studied nonobservance.

27. Which of the following statements about women/girls in non-industrial societies is supported by data from the *Standard Cross-Cultural Sample*?
    a. Women tend to have more political rights compared to men.
    b. Girl's are socialized in late childhood to become more aggressive.
    c. Girl's are trained to assume adult roles earlier than boys.
    d. Women have the most authority over older children.
    e. Women do more hunting than gathering.

28. According to the study by DeLoach, Cassidy and Carpenter, mothers tend to identify gender-neutral characters in children's stories as
    a. female.
    b. male.
    c. neither male or female.
    d. females when reading to their daughters, but male when reading to their sons.
    e. non-human.

29. Children are presented with a picture of a bear taking a group of young bears to a playground. The children are expected to identify the older bear as female. This is an example of
    a. a sex-typed activity.
    b. a feral bear.
    c. instinctive behavior.
    d. impression management among bears.
    e. universal gender.

30. Richer observed that sex-segregated play is primarily learned from
    a. teachers.
    b. the principal.
    c. parents.
    d. older children.
    e. younger children.

True/False

Answers and page references are provided at the chapter end.

1. The socialization process is primarily limited to early child rearing practices.

2. Contemporary studies indicate that researchers have been unable to replicate the highly publicized "Mozart Effect."

3. The rule of object permanence is discovered when a child is in Piaget's preoperational stage of cognitive development.

4. Chomsky argues that the rules of Universal Grammar are inborn.

5. Pidgin is to creole as jargon is to a complex language.

6. Bain observed that young children discover and use self-related words before they employ other-related words.

7. Margaret Mead maintained that sex roles are biologically determined and could be shaped to take any form.

8. Kohn's research indicates that parents who are employed in highly supervised work environments tend to encourage their children to conform to the expectations of others.

9. Parents encountering gender-neutral characters in children's stories tend to identify the characters as females since women are generally seen as being responsible for child care.

10. Richer observed that sex-segregated play is more prevalent among older children aged 6-14 than younger children aged 3-4.

Short Answer Questions

These short answer questions are provided to test your knowledge and understanding of the basic sociological concepts presented in Chapter 6. Page references for answers are included at the chapter end.

1. What do the case studies of feral children teach us about the nature of the socialization process?

2. What is the "Mozart Effect" and is the phenomenon supported scientifically?

3. Identify Piaget's four stages of cognitive development.

4. According to Chomsky what are three key aspects of language?

5. How does the development of the Nicaraguan Sign Language among deaf children illustrate the creation of a pidgin and a creole?

6. What conclusions did Margaret Mead reach concerning the sex-role socialization patterns of the Arapesh and Mundugumor of New Guinea?

7. Provide several examples of differential socialization.

8. Why do researchers conduct longitudinal studies?

9. Provide an example of studied nonobservance and note how it enhances role performance.

10. Identify several basic gender role and gender socialization trends in non-industrial societies.

Essay Questions

These questions are designed to test your understanding of key sociological concepts presented in Chapter 6 and your ability to apply these insights to concrete situations.

1. An increasing number of older persons are being placed in nursing homes. How does the quality of care received impact the continued socialization of the residents?

2. Compare and contrast stimulus – response (S – R) theory and Piaget's theory of cognitive stages.

3. Researchers argue that language is learned but also may be instinctive. Provide evidence supporting each claim.

4. How is early childhood socialization influenced by environmental factors and physiological constraints?

5. Researchers are starting to reconsider the impact of birth order effects on human behavior. Explain how birth order is an example of differential socialization.

Answers

Multiple Choice

1. c (149)
2. e (150)
3. a (151)
4. e (152-153)
5. e (153)
6. d (154-155)
7. b (154)
8. c (154)
9. d (155)
10. c (156)
11. d (156)
12. a (157)
13. b (158)
14. c (159)
15. e (160)
16. b (160)
17. e (160-162)
18. c (162-163)
19. a (162)
20. b (163-164)
21. c (165)
22. d (165-166)
23. a (166)
24. b (166)
25. c (168)
26. e (168-169)
27. c (170)
28. b (171-172)
29. a (171)
30. d (172-173)

True/False

1. F (150)
2. T (153)
3. F (154)
4. T (156)
5. T (157)
6. F (159)
7. F (161-162)
8. T (165-166)
9. F (171-172)
10. T (172-173)

Short Answer

1. (149-152)
2. (152-153)
3. (154-155)
4. (156)
5. (157-158)
6. (161-162)
7. (163-165)
8. (166)
9. (168-169)
10. (169-170)

CHAPTER 7

CRIME AND DEVIANCE

Extended Abstract

Norms regulate and guide behavior, but norms are not always followed. Sociologists refer to the violation of widely shared, clearly stated norms as acts of deviance. Since norms vary from group to group and society to society, deviance is a relative concept. Crimes are more formal acts of deviance. Crimes are behaviors that are prohibited by law. Since this definition of crime suggests that crimes are subject to varying political and legal interpretations, crime is defined as forceful or fraudulent acts that promote self-interest. Robbery, burglary and homicide are cited as examples of ordinary crime. These ordinary crimes are often unplanned, spontaneous and yield small rewards.

Various theories of deviance and crime are introduced in this chapter. They may be grouped under the following headings: biological theories, psychological theories and sociological theories. The biological theories suggest that deviance is genetically based. Lombroso, who maintained that some people were "born criminals," offered a classic biological theory. According to Lombroso criminals could be recognized according to certain physical traits, like low foreheads, and their genetic make-up predisposed them to deviant acts. Criminals should therefore be removed from society and imprisoned. Subsequent research by Goring and Pearson revealed that many of the physical characteristics attributed to the criminal population by Lombroso were not specific to the criminal population.

More recent biological studies within a behavioral genetics context have addressed deviance patterns among identical and fraternal twins. Higher rates of deviance have been observed among identical twins, twins who share the same genetic background. An interesting study by Gove looks at the interaction of gender, age, changing physiology and deviance. Generally, males are more likely to commit crimes than females, and they commit different types of crime. Also, crimes tend to decline with age. Why? For Gove the key is changing human physiology. Physical strength and energy peak for persons in their twenties and testosterone and adrenaline levels decline. Thus, older persons and women are generally less likely to commit more physically demanding crimes.

Studies attempting to link deviance to specific personality traits have yielded contradictory findings although there does appear to be a correlation between mental illness and violent crime. Gottfredson and Hirschi offer a personality theory that links crime weak self-control. Supposedly persons with weak self-control are unwilling to delay need satisfaction. They are thrill seekers who are self-centered and look for easy, quick solutions to problems. Criminal acts offer a quick, easy, enjoyable solution to obtaining desired needs.

The remainder of Chapter 7 is devoted to introducing diverse sociological theories of deviance. Two theories of deviant attachment are offered (differential association and subcultural deviance), in addition to a stratification theory (structure strain), a utility theory (control theory), a social bond theory (anomie theory) and a power theory (labeling theory). In developing the theory of differential association, Sutherland links deviance to the forming of deviant attachments. People commit deviant acts as they conform to the expectations of family members and friends who are deviant. The theory does not adequately address how deviant family members and friends became deviant initially. The subcultural theory of deviance focuses on whose norms become part of a society's legal code. Powerful groups establish laws that define what is normative and deviant.

Structure strain theories of deviance attribute deviance to experiences of inequality and unfavorable social position. Socially disadvantaged persons do not have access to the resources needed to obtain society's rewards, like wealth. Consequently, disadvantaged persons use deviant or illegal means to achieve their desired goals. Merton argues that crime is a result of poverty. However, there are problems with this theory also. The majority of poor persons do not commit deviant acts, and crime does not vary significantly by social class. More affluent persons commit a significant number of crimes like embezzlement. These crimes are known as white-collar crimes. On the other hand, crime rates can be high in poorer areas because many poor areas include mixed-use neighborhoods. These neighborhoods provide more opportunity for deviance by placing retail establishments and residential units in close proximity.

The control theory of deviance assumes that deviant acts are attractive. This theory attempts to explain why people conform rather than explaining why they commit deviant acts. People will conform when they gain more from acts of conformity than from deviance. When persons possess a high stake in conformity, social ties are strong, and deviant acts can destroy these bonds. Social ties that are risked through deviant acts include attachments (family, friends), investments (career), involvements (time and energy) and beliefs (self-respect). Combining aspects of differential association and control theory, Linden and Fillmore maintain that persons with low stakes in conformity risk little if they form deviant attachments. These factors interact and increase the tendency to commit deviant acts.

The anomie theory of deviance links deviance to the degree of social and moral integration experienced in a community. Social integration addresses the strength of social ties and bonds; whereas, moral integration addresses shared belief systems. Deviance will be low in areas where social and moral integration is high. Consequently, a community experiencing little population turnover and a high degree of religious involvement should experience a lower incidence of deviance.

Environmental factors such as climate and seasonal variations appear to correlate with crime. Crimes are higher during the summer months and the December holiday season. Pleasant weather conditions and holiday shopping provide contexts whereby crimes can be committed more easily.

The last sociological theory of deviance presented is labeling theory. This theory suggests that deviance results from having been socially defined as deviant. Persons are stigmatized and conform to the image (label) portrayed by the stigma. Therefore, how does society determine what is normative and what is deviant? Who has the power to resist being labeled? How do labels impact occupational opportunities, the development of interpersonal networks and a person's self-image? Labeling theory seeks to answer these questions and provides an explanation for why people commit repeated offenses, but it does not provide an adequate rationale for why people initially commit deviant acts.

A discussion of drugs and crime is introduced next. Social scientists maintain that alcohol and drug use stimulate crime and that persons will commit deviant acts to support habits. In a recent study conducted by the National Institute of Justice, urine samples were collected from persons residing in selected U.S. cities who were arrested over a two-week period. Drug use was high among the persons arrested and was higher for females. Although this does not mean that drug use caused the deviant acts, drug use appears to be a characteristic of an offender's life style. Stark concludes this chapter by noting important interrelationships among the theories presented and suggests how each may contribute to the development of a general theory of deviance.

Key Learning Objectives

After a thorough reading of Chapter 7, you should be able to:

1. distinguish crime and deviance and identify the problems associated with providing a clear definition of crime.

2. note major demographic trends associated with the incidence of ordinary crime like robbery, burglary and homicide.

3. identify major characteristics associated with the "typical" criminal act.

4. critically evaluate classical and contemporary biological theories of deviance.

5. state how Gove draws insight from contemporary biological and social-scientific research to portray the interaction among age, gender, changing human physiology and deviance.

6. explain why an increasing number of persons who are arrested for criminal acts also suffer from mental illness.

7. clarify the association between gratification, self-control and deviance.

8. identify the strengths and weaknesses associated with the two theories of deviant attachment (differential association and the subcultural theory of deviance).

9. evaluate the structural strain theory of deviance and clarify the association between social class, deviance and "white-collar" crime.

10. identify how attachments, investments, involvements and beliefs impact one's stake in conformity.

11. describe the interaction between low stake in conformity, deviant attachments and crime.

12. state the association among social integration, moral integration and deviance.

13. explain how climate change and seasonal shopping patterns could influence crime rates.

14. critically evaluate the labeling theory of deviance.

15. note important recent findings on the nature of the association between drug use and crime.

Chapter Outline

I. Deviance, Crime and the Criminal Act
    A. Deviance: The Violation of Widely Shared, Clearly Stated Norms
    B. Crime: A Problem of Definition
        1. Crime defined as behavior prohibited by law.
        2. Crime defined as acts of force and fraud promoting self-interest.
    C. Ordinary Crime: Robbery, Burglary and Homicide
        1. More than half of robberies occur on the street and are unplanned.
        2. Burglaries take place close to home and are most often committed by white males under 25.
        3. Homicide victims and offenders are often of the same sex, race and age.
    D. Characteristics of the "Typical" Criminal Act
        1. Crimes are generally unplanned and short-lived.
        2. Criminal acts are exciting and offer immediate rewards.

II. Biological Theories of Deviance
    A. Lombroso's Classical Theory
        1. Persons are born criminals.
        2. The criminal population should be removed from society.
        3. The criminal population possesses distinct physical characteristics.
        4. Goring and Pearson falsify several of Lombroso's claims about criminals' physical characteristics.
    B. Behavioral Genetics: Twin Studies and Deviance
        1. When an identical twin commits a crime, the odds are 50-50 that the other twin will be involved in criminal activity.
        2. When a fraternal twin commits a crime, there is only a 1 in 5 chance that the other twin will be involved in criminal activity.
    C. Gove: A Biological-Sociological Synthesis
        1. Crime rates vary by age and gender.
        2. Physical strength, energy, testosterone and adrenaline levels decline with age.
        3. Physically demanding and aggressive crimes decline with age.

III. Mental Illness, Personality and Deviance
    A. New Link between Mental Illness and Crime
        1. Increase in arrest of persons who suffer from mental illness.
        2. Increase attributed to changes in commitment laws and treatment policies.
    B. Deviance and Personality Traits: Conflicting Findings
    C. A Possible Link Between Deviance and Low Self-Control
        1. Persons committing criminal acts are less willing to delay gratification of needs.

2. Persons committing criminal acts are self-centered, thrill seekers looking for a simple, easy means to meet their needs.

IV. Sociological Theories of Deviance

A. Theories of Deviant Attachment

1. Differential association theory states that persons form strong social ties with family members and friends involved in deviant acts.
2. According to the subcultural theory of deviance, conflict and power struggles determine whose norms become part of a society's legal code and which acts will be classified as normative and deviant.

B. A Stratification Theory: Structural Strain Theory

1. Crime is a product of inequality and social disadvantage.
2. Crime results when acceptable means for achieving goals are unavailable.
3. Studies suggest that frequency of crime does not vary by social class.
4. Higher status persons commit different types of crime (white-collar crime).

C. Explaining Why People Conform: Control Theory

1. People conform when their stake in conformity is high and there is much to lose.
2. The stake in conformity is influenced by attachments, investments, involvements and beliefs.
3. Linden and Fillmore argue that persons with low stakes in conformity tend to form deviant attachments.

D. Community Stability and Deviance: Anomie Theory

1. Deviance increases as the number and importance of social attachments (social integration) decreases.
2. Deviance increases as agreement on societal norms (moral integration) decreases.

E. Deviance and Environmental Change

1. Opportunities for crime increase in more pleasurable climates, as people are more willing to travel and leave home.
2. Holiday shopping seasons offer more opportunities for crime.

F. Stigmatization and Deviance: Labeling Theory

1. Acts that lead others to label persons as deviant are known as primary deviance.
2. Secondary deviance involves acts that are committed as a result of having been labeled deviant.
3. Deviant labels limit a person's occupational opportunities and range of attachments and can impact self-image.
4. Labeling theory provides an explanation for why crimes are repeated but offers little insight into why crimes are initially committed.

G. Drugs and Crime: New Findings

1. Social scientists have argued that drug use stimulates violent behavior and that persons will commit crimes to support drug habits.
2. Data from the recent National Institute of Justice study indicate that drug use is high among persons arrested and is higher for females arrested.

3. Although drug use may not directly cause crime, it does appear to be part of the offender life style.

Key Terms

Based on your reading of Chapter 7, you should be able to define and illustrate the key sociological concepts listed below. Page numbers are provided in parentheses as reference points.

Deviance (177)
Crime (178)
Robbery (18)
Burglary (181)
Homicide (182)
Offender Versatility (183)
Born Criminals (184)
Self-Control (189)
Differential Association Theory (190)
Subcultural Deviance (192)
Structure Strain (193)
Structural Strain Theories (193)
White-Collar Crime (195)
Violations of Trust (196)
Control Theory (196)
Stake in Conformity (196-197)
Social Bonds (197)
Attachments (198)
Investments (199)
Involvement (199)
Beliefs (199-200)
Internalization of Norms (200)
Anomie (202)
Moral Communities (202)
Social Integration (202)
Moral Integration (202)
Labeling Theory (204)
Primary Deviance (204)
Secondary Deviance (204)

Key Research Studies

Listed below are key research studies cited in Chapter 7. Familiarize yourself with the major finding of these studies. Page references are provided in parentheses.

Gottfredson and Hirschi: define crime as acts of force and fraud designed to enhance self-interest. Develop theory of deviance based on low self-control, immediate gratification and thrill seeking (178-179, 189-190).

Lombroso: people are born criminals. The criminal population is distinguished by certain physical features. Criminals should be removed from society (184-185).

Goring and Pearson: employed more sophisticated statistical techniques to refute some of Lombroso's claims about the deviant population (185).

Gove: investigated the interaction of age, gender, changing human physiology and deviance. Observed that more physically demanding crimes decline with age due to declines in physical strength, energy and testosterone levels (186-188).

Sutherland: developed theory of differential association. Deviance linked to the development of strong attachments with deviant individuals (190).

Merton: developed structural strain theory of deviance. Linked deviance to social position and experiences of inequality. Deviance increases when persons do not have access to acceptable means of achieving desired goals (193-194).

Linden and Fillmore: developed theory of deviance linking key ideas from control theory and differential association theory. Children with weak parental and school ties have a low stake in conformity and are more likely to establish deviant attachments and commit deviant acts (200-201).

Durkheim: linked deviance and societal integration. Argued that deviance would be higher in cities due to a weaker network of social ties and less agreement on a common body of norms (202).

Guerry: noted seasonal variations in early nineteenth century crime rates and climate patterns (202-203).

National Institute of Justice Study: study based on arrest records from major U.S. cities. Urine samples were taken to test for drug use among offenders. High rates of drug use among offenders were discovered indicating that drug use may be part of the offender life style (205-206).

Info Trac Search Words

Enter these search terms to conduct more extensive investigations of key topics introduced in Chapter 7.

Deviant Behavior
Subdivisions: Research

Hate Crimes
Subdivisions: Social Aspects
Select one of the studies cited. Identify the sociological perspective used to provide a clearer understanding of the nature of hate crimes.

Criminal Statistics
Subdivisions: Analysis

Crime and Inequality
Choose one of the studies listed. How is inequality measured? Is the study grounded in the structural strain perspective? State the study's major findings.

Drug Use and Crime

Multiple Choice

Answers and page references are provided at the chapter end.

1. Norm violations are classic examples of
   a. socialization.
   b. deviance.
   c. attachments.
   d. conformity.
   e. integration.

2. Gottfredson and Hirschi define crime as
   a. carefully planned acts of deviance.
   b. genetic predispositions toward violence.
   c. behaviors prohibited by law.
   d. forceful and fraudulent acts that promote self-interest.
   e. unacceptable acts primarily committed by middle-class and upper-class citizens.

3. Available crime data indicate that __________ likely to be arrested for robbery.
   a. males under age 25 are more
   b. females under age 25 are more
   c. females 25 and over are more
   d. males 25 and over are more
   e. males and females are equally

4. Criminal acts
   a. tend to be committed on the "spur of the moment."
   b. are brief in duration.
   c. offer immediate rewards.
   d. tend to be exciting.
   e. all of the above

5. Lombroso argued that persons are "born criminals." This is an example of a/an __________ theory of deviance.
   a. biological
   b. sociological
   c. psychological
   d. cultural
   e. economic

6. Lombroso portrayed criminals as
   a. being very strong.
   b. having small jaws.
   c. being intelligent.
   d. being small in stature.
   e. none of the above

7. Which of the following statements is true?
   a. Females are more likely than males to be arrested.
   b. Fraternal twins are more likely to commit deviant acts than are identical twins.
   c. The criminal records of adopted children more closely resemble the criminal records of their biological rather than their adoptive parents.
   d. Compared to younger persons, older persons are more likely to be arrested.
   e. Each statement is false.

8. Data from the U.S. Department of Justice indicate that women are least likely to be arrested for
   a. embezzlement.
   b. robbery.
   c. fraud.
   d. runaways.
   e. forgery.

9. Gove's research indicates that people are
   a. less likely to commit physically demanding crimes as they age.
   b. more likely to commit physically demanding crimes as they age.
   c. equally likely to commit physically demanding crimes as they age.
   d. more likely to report being victimized by crime as they age.
   e. less likely to report being victimized by crime as they age.

10. The increase in the arrest of persons who are mentally ill has been attributed to such factors as
   a. changes in commitment laws.
   b. changes in treatment policies.
   c. severe mental hospital overcrowding.
   d. a and b
   e. b and c

11. In developing their theory of deviance and self-control, Gottfredson and Hirschi maintain that
   a. higher rates of deviance are associated with high rates of self-control.
   b. persons committing deviant acts tend to think more about others' interests and concerns.
   c. persons committing deviant acts are risk takers and thrill seekers.
   d. persons commit deviant acts because they like to delay need gratification
   e. deviant acts are carefully planned and involve complex performance strategies.

12. This theory of deviance could also be called the "guilt by association theory."
   a. control theory
   b. differential association theory
   c. self-control theory of deviance
   d. anomie theory
   e. labeling theory

13. According to this theory, the more powerful groups in a society define which acts are deviant and which are normative.
    a. differential association theory
    b. anomie theory
    c. control theory
    d. structural strain
    e. subcultural theory of deviance

14. Social class is believed to be a key factor influencing deviance according to this theory of deviance.
    a. control theory
    b. anomie theory
    c. differential association theory
    d. structural strain theory
    e. self-control theory

15. *General Social Survey* data indicate that the percent of people who have been "picked up" by the police is highest among persons who viewed their family income as
    a. far below average when they were younger.
    b. below average when they were younger.
    c. average when they were younger.
    d. above average when they were younger.
    e. the percentages are roughly equal for all income groups.

16. Crime generally committed by more affluent members of society is labeled
    a. white-collar crime.
    b. blue-collar crime.
    c. pink-collar crime.
    d. red-collar crime.
    e. black-collar crime.

17. The "stake in conformity" is primarily associated with this theory of deviance.
    a. labeling theory
    b. structural strain
    c. control theory
    d. subcultural theory
    e. anomie theory

18. Control theorists link conformity to strong social bonds. If a person is convicted of a deviant act that person could lose their job. The social bond that could be endangered is
    a. attachments.
    b. beliefs.
    c. investments.
    d. involvements.
    e. self-control.

19. *General Social Survey* data reveal that arrest rates are lowest among
    a. married persons.
    b. single persons.
    c. divorced persons.
    d. separated persons.
    e. arrest patterns do not vary by marital status.

20. According to control theorists, the amount of time and energy allocated to acts of conformity is known as
    a. beliefs.
    b. involvements.
    c. investments.
    d. attachments.
    e. integrations.

21. National survey data indicate that arrest rates are highest among people who attend church
    a. daily.
    b. weekly.
    c. monthly.
    d. yearly.
    e. less than yearly.

22. Linden and Fillmore propose that
    a. primary deviance leads to secondary deviance.
    b. secondary deviance leads to primary deviance.
    c. persons with high stakes in conformity tend to form delinquent attachments.
    d. persons with low stakes in conformity tend to form delinquent attachments.
    e. high degrees of moral and social integration are associated with high rates of deviance.

23. When society's norms are not clear or are no longer followed, a condition of normlessness exists. Durkheim labeled this phenomenon
    a. moral communities.
    b. anomie.
    c. social integration.
    d. deviance.
    e. stake in conformity.

24. Social integration measures
    a. the degree of agreement on society's norms.
    b. the degree of stratification present in society.
    c. the strength of bonds and attachments.
    d. the rate of population growth.
    e. personal self-control.

25. According to anomie theory, deviance increases as social integration
  a. and moral integration increase.
  b. increases and moral integration decreases.
  c. decreases and moral integration increases.
  d. and moral integration decrease.
  e. none of the above

26. Researchers consistently note that crime rates tend to be higher during the __________ months.
  a. fall
  b. winter
  c. spring
  d. summer
  e. they remain high throughout the year

27. The distinction between primary and secondary deviance is a critical feature of this theory of deviance.
  a. anomie theory
  b. control theory
  c. subcultural theory
  d. structural strain theory
  e. labeling theory

28. Being labeled deviant may
  a. limit a person's economic opportunities.
  b. restrict a person's occupational opportunities.
  c. limit a persons ties with more conventional people.
  d. impact one's self-image in a negative way.
  e. all the above

29. Recent research on drug use indicates that the most widely used drug among the U.S. population is __________, but among persons arrested it is __________.
  a. cocaine; heroin
  b. heroin; LSD
  c. marijuana; cocaine
  d. LSD; marijuana
  e. cocaine; LSD

30. The study on offender drug use conducted by the National Institute of Justice measured drug use by
  a. taking urine samples.
  b. offering offenders reduced sentences if they told the truth about drug use.
  c. administering a polygraph test.
  d. having offenders complete a questionnaire.
  e. administering an EKG.

True/False

Answers and page references are provided at the chapter end.

1. More than half of all robberies occur on the street and most are unplanned.

2. Most criminal acts are viewed as exciting, are brief in duration and offer immediate rewards.

3. If one twin is involved in criminal behavior, the odds are higher that the second twin will be involved in criminal activities if the twins are identical rather than fraternal twins.

4. Gove discovered that physical strength and energy increases significantly after age 30. This explains why older persons are still able to commit physically demanding crimes.

5. Persons with high self-control seek immediate gratification and thus are more likely to commit deviant acts.

6. According to Sutherland's theory of differential association, one's risk of committing deviant acts increases as social ties with deviant persons become stronger.

7. White-collar crimes are committed primarily by persons from the lower class.

8. According to the control theory of deviance, deviance increases as one's stake in conformity decreases.

9. Anomie theory predicts that higher rates of crime will be experienced in societies where the norms are strong and widely shared.

10. Labeling theorists define primary deviance as behavior that results from having been labeled deviant.

Short Answer Questions

These short answer questions are provided to test your knowledge and understanding of the basic sociological concepts presented in Chapter 7. Page references for answers are included at the chapter end.

1. What research problems are presented when crime is defined as behavior prohibited by law?

2. Why would robbery, burglary and homicide be included as examples of ordinary crime?

3. According to Gove what factors contribute to the decline in crime rates among persons aged 30 and over?

4. What is the nature of the association between self-control and criminal acts?

5. State the strengths and weaknesses of the subcultural theory of deviance.

6. Why is structural strain theory an example of a stratification theory of deviance?

7. Provide examples of white-collar crime.

8. How does the degree of social and moral integration impact deviance?

9. How do labeling theorists distinguish primary and secondary deviance?

10. What are the major findings of the recent National Institute of Justice survey on drug use and crime?

Essay Questions

These questions are designed to test your understanding of key sociological concepts presented in Chapter 7 and your ability to apply these insights to concrete situations.

1. Explain why sociologists generally treat deviance and crime as relative concepts? Justify your answer with concrete examples.

2. Many urban planners favor mixed-use development projects. How might these development projects stimulate crime?

3. Develop an explanation for white-collar crime utilizing the deviant attachment and control theory perspective.

4. Provide a critical response to the following assertion, "people commit deviant acts because they are attractive."

5. Certain areas of the South have experienced substantial population growth over the past two decades. According to the anomie perspective, how would this social fact impact Southern crime rates?

Answers

Multiple Choice

1. b (177)
2. d (178)
3. a (180)
4. e (184)
5. a (184)
6. e (185)
7. c (185)
8. b (187)
9. a (187-188)
10. d (188)
11. c (189-190)
12. b (190-191)
13. e (192-193)
14. d (193-194)
15. e (194)
16. a (195-196)
17. c (196-197)
18. c (197, 199)
19. a (199)
20. b (199)
21. e (200)
22. d (200-201)
23. b (202)
24. c (202)
25. d (202)
26. d (203)
27. e (204)
28. e (204)
29. c (205)
30. a (205)

True/False

1. T (180)
2. T (184)
3. T (185)
4. F (187-188)
5. F (190)
6. T (190-191)
7. F (195)
8. T (196-197)
9. F (202)
10. F (204)

Short Answer

1. (177-179)
2. (179-182)
3. (187-188)
4. (189-190)
5. (192-193)
6. (193-195)
7. (195-196)
8. (202)
9. (204)
10. (2085-206)

CHAPTER 8

SOCIAL CONTROL

Extended Abstract

The rules of acceptable and unacceptable behavior are stated in a society's system of norms. Societies employ informal and formal means of social control to encourage conformity to societal expectations. Informal means of social control are expressed in the expectations and evaluations of family, friends, employers and role models. When informal means of social control fail to achieve the necessary level of conformity, formal mechanisms may be employed. In industrial societies, formal mechanisms of social control include the police, the courts and the prison system. Various informal and formal social control mechanisms are evaluated throughout this chapter.

The first example presented is the system of informal social control within Japanese society. The low rates of deviance in Japan are a well-known fact. What factors contribute to this? Research by Hechter and Kanazawa indicates that deviance is low because Japanese society is organized to encourage high levels of group conformity. Parents, teachers and employers closely supervise individuals and demand high levels of performance. Groups, like the family, school and work environments, exert a high degree of control over their members. This high degree of informal social control is maintained through the principles of dependence, visibility and extensiveness.

The principle of dependence states that individual conformity increases as persons become more dependent on a group. In Japan a teacher's or an employer's recommendation plays a critical role in determining a person's future success. Individuals discover early that being obedient is advantageous. The second principle, visibility, states that conformity increases when individual behavior is easily observed. At home, children are closely supervised, and at school, students are expected to remain in the classroom. Individual activities are discouraged. In the work place, many Japanese workers perform their duties in open office spaces. Each of these environments reinforces the public nature of Japanese daily life. The third principle is the principle of extensiveness. Here conformity increases as institutions have greater influence on more areas of one's life. In Japan teachers and employers do not restrict their influence to school and work related issues. Issues in one's private life are addressed openly and life style expectations are stated clearly. These informal mechanisms of social control interact to create and maintain a high conformity - low deviance environment.

The police, the court system and prisons represent formal mechanisms of social control. Formal mechanisms of social control seek to prevent crime by limiting opportunities for crime, deterring deviant acts through fear of punishment or reducing the incidence of deviance through reform programs. Stark evaluates each option.

Prevention programs assume that deviance is reduced as opportunities for deviance are reduced. According to the opportunity theory of deviance, crime increases when people are motivated to commit an offense, suitable targets are readily available and environments are not carefully monitored or guarded. Homes are more vulnerable to burglary since people generally are not home during the day. On the other hand, home burglar alarm systems represent an attempt to maintain a monitored environment. It is argued that early intervention programs may reduce the opportunity for deviance.

The Cambridge-Somerville experiment is an example of a classic intervention program. In this study young boys from economically distressed communities were randomly placed in experimental and control groups. Boys placed in the experimental group were provided with many of the educational and social opportunities that young boys from more affluent families would receive. The boys in the control group were not exposed to these benefits. It was hypothesized that the boys in the experimental group would be more likely to stay out of trouble as they grew older. At the completion of the study, researchers observed that the deviance rates for both groups were equivalent! This finding and similar findings challenge whether prevention programs are effective formal mechanisms of social control. However, some success has been observed among intervention programs that involve parents.

Many view punishment as a crime deterrent. One highly controversial deterrence mechanism is capital punishment. The effectiveness of deterrence as a mechanism of formal social control is evaluated by looking at Gibbs' theory of deterrence and Ehrlich's study on capital punishment. Gibbs argues that crime rates are lower when punishment is perceived as more certain, severe and swift. The threat of punishment must be perceived as real for punishment to function as a deterrent. Challenging a common perception among social scientists regarding the effectiveness of capital punishment as a crime deterrent, Ehrlich analyzed available homicide and execution data for the 1933-1969 period. The correlation between homicide and execution was negative and strong. Subsequent studies have produced mixed findings.

The last mechanism of formal control discussed is the criminal justice system. Major components of this system are the police, the courts and prisons. Stark estimates that only four in ten crimes is actually reported to the police and arrest rates vary significantly by type of crime. These findings suggest that the odds of being caught and punished for a crime are low. Thus, jail may not be perceived as a strong deterrent. Also, since many court cases are delayed, dismissed or plea-bargained, the certainty, severity and swiftness of punishment is questioned. Given these conditions it is not surprising that recidivism (repeat offense) rates are high. Finally, are prisons effective mechanisms of reform and resocialization?

Prisons are a relatively new phenomenon having been introduced in the U.S. by the Quakers in the late eighteenth century. Under the prison system, confinement and reform replace physical punishment as a more effective means of instilling conformity and maintaining social control. But are prison reform programs effective? Are prisoners adequately prepared to function in society following their release from prison? Labeling theory suggests that ex-convicts are stigmatized. Consequently their social attachments and economic opportunities are limited. The TARP (Transitional Aid Research Project) experiment attempted to determine whether economic assistance would enable released prisoners to complete the transition from prison back to society successfully. Prisoners in the experimental group received a weekly paycheck for six months while prisoners in the control group followed standard release procedures. It was hypothesized that the recidivism rate among the experimental group members would be low. Recidivism rates were found to be identical for the experimental and the control group. Many therapeutic prison reform programs have been introduced, but recidivism rates have remained high.

Key Learning Objectives

After a thorough reading of Chapter 8, you should be able to:

1. identify examples of informal and formal mechanisms of social control.
2. explain how group pressure can be an effective informal mechanism of social control.
3. note how dependence, visibility and extensiveness function as informal mechanisms of social control.
4. distinguish prevention, deterrence and reform as three principle mechanisms of formal social control.
5. outline the major features of Cohen and Felson's opportunity theory of deviance.
6. critically evaluate the Cambridge-Somerville delinquency prevention program.
7. identify important factors contributing to the failure of delinquency prevention – reform programs.
8. summarize the main tenets of Gibbs' theory of deterrence.
9. critically evaluate social-scientific studies addressing capital punishment's deterrent effects.
10. identify major trends in the data available on the reporting of crimes, arrests rates and confidence in the police.
11. specify features of the U.S. court system, which promote delays in punishment.
12. identify factors that could be associated with high recidivism rates.
13. outline the development of the U.S. prison system.
14. explain how ex-convicts may be stigmatized.
15. evaluate the effectiveness of prison reform and resocialization programs.

Chapter Outline

I. Informal Mechanisms of Social Control
   A. Family and Friend Networks and Conformity
   B. Solidarity and Conformity in Japan: Hechter and Kanazawa
      1. The principle of dependence states that group pressure increases conformity.

2. The principle of visibility states that observation by others enhances conformity.
3. The principle of effectiveness stresses that conformity increases when groups are granted legitimate authority to comment on personal life.

II. Formal Mechanisms of Social Control
   A. Prevention: Limiting Opportunities for Crime
      1. The Cohen and Felson opportunity theory of deviance.
         a. Persons must be motivated to commit crimes.
         b. Suitable crime targets must be available.
         c. Effective guardians are not available.
      2. Delinquency prevention programs like the Cambridge-Somerville experiment have limited impact.
         a. Boys from disadvantaged communities will be less likely to commit deviant acts if afforded the same economic and social opportunities as more affluent children.
         b. No difference in the delinquency rate of the experimental and control group was observed.
         c. Counselors do not impact behavior as strongly as do parents and friends.
   B. Deterrence: The Fear of Punishment
      1. Gibbs' Theory of Deterrence.
         a. Punishment must be perceived as rapid and swift in order to be an effective deterrent.
         b. To deter crime punishment must be perceived as certain.
         c. If punishment is perceived as severe, it will be a more effective deterrent.
      2. The Capital Punishment Controversy
         a. Early social-scientific research suggested that capital punishment is not a deterrent.
         b. Ehrlich discovers strong negative association between homicide and execution rates for the 1933-1969 period.
         c. Contemporary social-scientific research on the deterrent effect of capital punishment yields mixed results.
   C. The Criminal Justice System and Recidivism
      1. Only one in three crimes are reported to the police.
      2. Dismissed and plea-bargained cases reduce the risk of swift, certain and severe punishment.
      3. The present U.S. recidivism rate stands at 70 percent.
      4. Prisons were introduced as a more humane form of punishment.
         a. Physical punishment is replaced by prison reform and resocialization programs.
         b. Stigmatization limits ex-convicts' economic opportunities.
         c. The TARP experiment.
            i. Released prisoners were provided with economic assistance to ease their transition back to society.

ii. Ex-convicts receiving economic assistance (experimental group) were hypothesized to have lower recidivism rates.

iii. Experimental and control group recidivism rates were equal.

d. Available research fails to demonstrate that prison reform and resocialization programs reduce recidivism rates.

Key Terms

Based on your reading of Chapter 8, you should be able to define and illustrate the key sociological concepts listed below. Page numbers are provided in parentheses as reference points.

Social Control (211)
Informal Social Control (211)
Formal Social Control (211, 214)
Social Order (212)
Principle of Dependence (213)
Principle of Visibility (213)
Principle of Extensiveness (213)
Prevention (215)
Opportunity Theory (215)
Treatment Group (216)
Deterrence (218)
Capital Punishment (219)
Deterrence Theory (220)
Recidivism Rate (226)
Penitentiary (226)
Resocialization (227)

Key Research Studies

Listed below are key research studies cited in Chapter 8. Familiarize yourself with the major findings of these studies. Page references are provided in parentheses.

Hechter and Kanazawa: studied the high level of social control in Japanese society. Conformity is linked to influence of powerful group networks at home, school and work (212-214).

Cohen and Felson: developed an opportunity theory of deviance. Deviance is linked to motivation, available targets and weak observation mechanisms (215).

Cambridge-Somerville Experiment (Cabot, Powers and Witmer): unsuccessful delinquency prevention program designed to reduce delinquency among boys from disadvantaged communities. Boys in the experimental group were provided with economic and social experiences generally available only to the more affluent. Delinquency rates for the experimental and control groups were the same (215-218).

Gibbs: developed a deterrence theory linking crime reduction to the perceived swiftness, certainty and severity of punishment (220-221).

Ehrlich: analyzed homicide and execution data for the 1933-1969 period and observed a strong negative correlation between homicide and execution rates (222).

Auburn Prison: an early nineteenth century prison that provided a basis for modern prison design. Design featured highly visible cells, and dangerous criminals were placed in solitary confinement (226).

Lenihan, Rossi and Berk and the Transitional Aid Research Project (TARP): a prison reform and resocialization program designed to ease the transition of ex-convicts back to society. Ex-convicts in the experimental group received monetary assistance for a six-month period. Study revealed that recidivism rates for the experimental and control group were identical (227-228).

Info Trac Search Words

Enter these search terms to conduct more extensive investigations of key topics introduced in Chapter 8.

Social Control
    Subdivisions: Research

Opportunity Theory

Punishment in Crime Deterrence
    Subdivisions: Analysis
    Select one of the studies listed. How is punishment being presented or evaluated as a deterrent?

Recidivism
    Choose one of the studies on recidivism listed. Does the study provide any evidence of a reduction in recidivism rates? Are any factors associated with either an increase or decrease in recidivism rates identified?

Prisons
    Subdivisions: Confinement Conditions

Multiple Choice

Answers and page references are provided at the chapter end.

1. The term "highwaymen" refers to
    a. politicians who favor public transportation programs.
    b. police officers.
    c. persons who transport illegal goods.
    d. prison inmates assigned to work on highways.
    e. a singing group.

2. Collective efforts to increase conformity to society's norms and reduce deviance are known as forms of
   a. recidivism.
   b. opportunity.
   c. social control.
   d. dependence.
   e. anomie.

3. The police, the courts and prisons are
   a. part of the criminal justice system.
   b. primary examples of informal means of social control.
   c. primary examples of formal means of social control.
   d. a and b
   e. a and c

4. Social order measures
   a. the extent of inequality present in a society.
   b. the degree to which citizens follow society's norms.
   c. the effectiveness of gender role socialization.
   d. the concentration of power among political groups.
   e. all of the above

5. Hechter and Kanazawa argue that conformity to group norms increases as persons rely more on group members. This is known as the principle of
   a. dependence.
   b. resocialization.
   c. visibility.
   d. deterrence.
   e. extensiveness.

6. Japanese society is characterized by a high degree of conformity to social norms. Each social control practice is correctly stated except
   a. student job applications must be accompanied by a nomination from their school.
   b. high schools determine whether or not a student will attend college.
   c. students are closely supervised and individual activities are strongly discouraged.
   d. most Japanese work in open office areas so that their work may be observed easily by supervisors.
   e. each practice is stated correctly.

7. Prevention, deterrence and reform are all attempts to maintain
   a. deviance.
   b. visibility.
   c. formal social control.
   d. informal recidivism.
   e. opportunity crime.

8. Motivation to commit crime, availability of suitable targets and absence of effective guardians are major concepts associated with
   a. deterrence theory.
   b. the opportunity theory of deviance.
   c. anomie theory of deviance.
   d. structural strain theory.
   e. the subcultural theory of deviance.

9. Neighborhood-watch programs and home security systems represent society's attempts to
   a. increase effective guardianship.
   b. reduce social control.
   c. increase the availability of suitable crime targets.
   d. motivate people to commit crime.
   e. reduce social order.

10. The Cambridge-Somerville experiment was an example of
    a. prison reform.
    b. police reform.
    c. a delinquency prevention program.
    d. court reform.
    e. capital punishment research.

11. A treatment group is another name for a __________ group.
    a. control
    b. independent
    c. dependent
    d. spurious
    e. experimental

12. An "eye for an eye" is an example of
    a. prison reform.
    b. resocialization.
    c. prevention.
    d. deterrence.
    e. stigmatization.

13. Execution is another name for
    a. a control group.
    b. an experimental group.
    c. capital punishment.
    d. recidivism.
    e. resocialization.

14. Gibbs' deterrence theory is grounded in __________ theories of deviance.
   a. social learning and control
   b. biological and psychological
   c. anomie and labeling
   d. structural strain and control
   e. psychological and anomie

15. Gibbs argues that fewer crimes will be committed when punishment is perceived as
   a. slow and uncertain.
   b. swift and unpredictable.
   c. slow and certain.
   d. swift and certain.
   e. none of the above

16. *Gallup Poll* data indicate support for capital punishment in the U.S.
   a. has remained at the same level over the past six decades.
   b. has increased steadily from the 1930s.
   c. has decreased steadily from the 1930s.
   d. increased sharply during the 1950s and 1960s.
   e. declined sharply during the 1950s and early 1960s.

17. This religious group was the first to introduce prison sentencing in the U.S. as a more humane substitute for execution.
   a. Catholics
   b. Baptists
   c. Quakers
   d. Jews
   e. Pentecostals

18. Ehrlich's analysis of homicide and execution data for the 1933-1969 period indicates that
   a. homicide rates cannot be measured effectively.
   b. execution rates cannot be measured effectively.
   c. homicide and execution rates are positively correlated.
   d. homicide and execution rates are not correlated.
   e. homicide and execution rates are negatively correlated.

19. Contemporary research on the deterrent effect of capital punishment
   a. provides overwhelming support for the deterrent effects of capital punishment.
   b. clearly demonstrates that capital punishment is not a deterrent.
   c. has generally provided mixed support for the deterrent effect of capital punishment.
   d. is highly suspect since adequate measures of effectiveness are not readily available.
   e. has yielded unacceptable findings because proper experimental and control group strategies have not been employed.

20. It is estimated that only __________ crimes are reported to the police.
   a. one in ten
   b. one in five
   c. two in five
   d. three in five
   e. seven in ten

21. The most frequently reported crime is
   a. robbery.
   b. auto theft.
   c. rape.
   d. assault.
   e. murder.

22. Data from the *World Values Survey* indicate that confidence in the police is highest among residents in
   a. Denmark.
   b. England.
   c. United States.
   d. Canada.
   e. Mexico.

23. In the U.S. a person is least likely to be arrested for committing
   a. auto theft.
   b. burglary.
   c. rape.
   d. assault.
   e. homicide.

24. Which of the following statements is false?
   a. Prosecutors drop many arrest charges because of flaws in arrest procedures.
   b. Sentences are often reduced through plea-bargaining.
   c. Many persons convicted of crimes receive suspended sentences or are placed on probation.
   d. Time off for good behavior can reduce a person's sentence by as much as 25 %.
   e. Each statement is true.

25. Present recidivism rates in the U.S. are approaching
   a. 25%.
   b. 50%.
   c. 70%.
   d. 90%.
   e. 100%

26. Imprisonment began to be seriously used as a more humane alternative to physical punishment
   a. in Roman times.
   b. during the Middle Ages.
   c. when the Spanish came to North America.
   d. during the eighteenth century.
   e. following World War II.

27. Currently the annual cost associated with keeping a person in prison is
   a. $10,000.
   b. $25,000.
   c. $40,000.
   d. $50,000.
   e. $75,000.

28. Today many prisons try to enhance social control among prisoners by
   a. executing more prisoners on a regular basis.
   b. placing prisoners in solitary confinement for six months when they first enter prison.
   c. administering high doses of electric shock when they break prison polices.
   d. requiring prisoners to raise money to pay for the cost of prison care.
   e. involving prisoners in therapeutic reform and resocialization programs.

29. In the TARP experiment, ex-convicts who received economic support to aid them in their transition from prison back to society were part of the __________ group.
   a. experimental
   b. independent
   c. spurious
   d. control
   e. dependent

30. The TARP study demonstrated that providing prisoners with economic support during their transition from prison to society
   a. only reduced the recidivism rates of male ex-convicts.
   b. only reduced the recidivism rates of female ex-convicts.
   c. generally increased ex-convict recidivism rates.
   d. had no impact on ex-convict recidivism rates.
   e. significantly reduced ex-convict recidivism rates and has now been introduced in most prisons nationally.

True/False

Answers and page references are provided at the chapter end.

1. The police, the courts and prisons are examples of formal mechanisms of social control.

2. The principle of extensiveness developed by Hechter and Kanazawa states that conformity increases as group members become more dependent on the group.

3. According to the opportunity theory of deviance, crime increases as the number of suitable targets increases. Therefore, crime rates are lower in areas that are farther away from fast-food restaurants.

4. The Cambridge-Somerville experiment is an example of an effective delinquency prevention program.

5. Delinquency prevention programs are successful because counselors strengthen weak network attachments and are adequate substitutes for parents and close friends.

6. Deterrence theory argues that conformity increases when punishment is perceived as certain and severe.

7. Catholics were the first group in the U.S. to introduce prison sentences as a more humane form of punishment.

8. Generally only four in ten crimes committed is reported to the police.

9. Recidivism rates in the U.S. are high. Approximately 70 % of persons convicted of a crime are likely to be convicted of another crime.

10. The TARP experiment demonstrated that ex-convicts who received financial help in making the transition back to society were less likely to become repeat offenders.

Short Answer Questions

These short answer questions are provided to test your knowledge and understanding of the basic sociological concepts presented in Chapter 8. Page references for answers are included at the chapter end.

1. Distinguish informal and formal means of social control by providing examples of each.

2. Hechter and Kanazawa identify three factors impacting the degree of social control that groups may exert over individuals. These factors are stated in terms of principles. Identify each principle.

3. According to opportunity theory, what conditions favor an increase in deviance?

4. Identify the main findings of the Cambridge-Somerville experiment.

5. Outline the major tenets of Gibbs' deterrence theory.

6. How do Ehrlich's findings on the homicide rate – execution rate association differ from previous social-scientific research on the association?

7. Which crimes are more likely to be reported to the police and which are least likely to be reported?

8. What factors influence the severity of sentences rendered by the courts?

9. How do modern therapeutic prisons differ from older punitive prisons?

10. The TARP study is an example of a prison resocialization program. How effective was the program in reducing recidivism rates?

Essay Questions

These questions are designed to test your understanding of key sociological concepts presented in Chapter 8 and your ability to apply these insights to concrete situations.

1. Apply Hechter and Kanazawa's principles of dependence, visibility and extensiveness to an analysis of informal mechanisms of control in U.S. society.

2. Explain seasonal variations in crime utilizing the opportunity theory perspective.

3. Present arguments for and against capital punishment as a crime deterrent.

4. Briefly outline the development of the prison system in the U.S.

5. Even though many prison resocialization programs have been introduced, they have been ineffective in reducing recidivism rates. Why have recidivism rates remained high?

Answers

Multiple Choice

1. d (211)
2. c (211)
3. e (211)
4. b (212-213)
5. a (213)
6. e (213-214)
7. c (215)
8. b (215)
9. a (215)
10. c (215-216)
11. e (216)
12. d (218)
13. c (219)
14. a (220)
15. d (220)
16. e (222)
17. c (222)
18. e (222)
19. c (223)
20. c (224)
21. b (223)
22. a (224)
23. b (225)
24. e (225-226)
25. c (227)
26. d (226)
27. d (227)
28. e (227)
29. a (228)
30. d (228)

True/False

1. T (211)
2. F (213)
3. T (215)
4. F (216-217)
5. F (218)
6. T (220)
7. F (222)
8. T (224)
9. T (226)
10. F (228)

Short Answer

1. (211-212)
2. (213-214)
3. (215)
4. (215-217)
5. (220-221)
6. (222)
7. (223-225)
8. (225-226)
9. (226-227)
10. (227-228)

CHAPTER 9

CONCEPTS AND THEORIES OF STRATIFICATION

Extended Abstract

Societies are characterized by varying degrees of inequality or stratification. Two major stratification concepts are social class and social mobility. In a class-based society people occupy positions of unequal rank; however, to the extent that opportunities for social mobility exist, movement between and within positions may be possible. Stark concludes this chapter by discussing theories of stratification from functionalist, evolutionary and conflict perspective.

Marx and Weber provide two classic definitions of social class. Marx maintained that inequality is rooted in the ownership of the means of production. The means of production includes anything that generates wealth such as land, machines and tools. In a capitalist society, two groups compete for the ownership of the means of production. These groups or classes are the bourgeoisie, the owners of the means of production, and the proletariat, the workers. Due to the urban bias of Marx's model, farmers were excluded from the two-class system. Also, migrant workers and criminals, referred to as the lumpenproletariat, were perceived as marginally active in the capitalist economy. The real source of conflict in society is between the bourgeoisie and the proletariat. Marx believed that revolution would occur as the proletariat attained class consciousness, an awareness of a common position, mutual interests and a common enemy. False consciousness would arise when workers assumed that owners also shared their interests.

Marx argued that economic factors, like property ownership, were the main determinants of social class. On the other hand, Weber maintained that stratification was based on class, status and party. Today these factors are identified as a group's unequal access to property, prestige and power. Property refers to the ownership or control of material possessions. Prestige is grounded in social perception and reflects the honor and power associated with one's position. Power involves the ability to issue a command and enforce it. Since the three dimensions of stratification may act independently, status inconsistency results. Weber argued that persons or groups could acquire significant material possessions and lack prestige and power. Conversely, prestige and power could lead to the further accumulation of power. Lenski argues that as people interact they present themselves in terms of their highest perceived rank, but they are responded to in terms of their lowest perceived rank. Different groups attempt to preserve their social position in this manner.

Although individuals and groups occupy positions of unequal rank within a stratified society, persons may be able to move within or between ranks. This movement is known as social mobility. Two factors influencing social mobility are achieved status and ascribed status. Achieved status is based on merit while ascribed status is assigned at birth and is fixed. Therefore, a college graduate would be an example of an achieved status; whereas, race and gender are ascribed statuses. Class systems tend to place more emphasis on achieved statuses, while caste systems place more emphasis on ascribed statuses. A greater degree of social mobility is allowed in class-based stratification systems. Also, a greater degree of upward or downward mobility may exist within a society if the distribution of available upper status or lower status positions changes. This is known as structural mobility, which is often contrasted with exchange mobility. When exchange mobility occurs, the distribution of upper and lower status positions is fixed. Some individuals or groups will move up only if others move down. Finally, classes are able to maintain their social position through the creation and maintenance of distinctive speech styles and tastes. This is known as cultural capital. Bourdieu observed that upper-class cultures stress abstract things, involvement in the arts and intellectual leisure activities while lower-class cultures stress concrete thought and focus on life's necessities. Stark suggests that upper-class groups are characterized more by cosmopolitan rather than local networks.

Marx argued that inequality would be reduced through the recreation of a classless society, a utopian society. This would take place if the private ownership of the means of production was replaced by state ownership. At this point everyone would be part of the proletariat. However, Dahrendorf has argued that inequality exists within the state because those who control the government have more power. Likewise, Mosca maintained that stratification is inevitable since political development is associated with differences in power. Those with power will exploit others in order to maintain their advantages.

Stark concludes the chapter with a discussion of stratification from three different sociological perspectives. The Davis and Moore functionalist theory of stratification is grounded in the concept of replaceability, the functional importance of a task and a system of unequal rewards. Persons can be motivated to fill important, demanding roles by offering greater rewards. Likewise, as a person's work becomes less replaceable, the functional importance of the task increases. Social and economic rewards are linked to a task's functional importance. The society is offered as an illustration of replaceability. Ay is the least replaceable member because Ay is the only member who can provide air. Inequality is then linked to replaceability. Evolutionary theory ties stratification to specialization. Over time society accumulates aspects of culture that are effective in addressing society's needs. But over time, no one can be an expert in everything. Areas of specialization develop, and some areas are regarded as more important than others. The result is stratification. According to the conflict perspective, stratification is associated with the exploitation and manipulation of replaceability. Persons in power tend to maximize their self-interests by exploiting the resources of others. Likewise, professional groups can restrict the number of people working in their field and control replaceability by relying on a system of credentials and expertise (meritocracy). Also, since unions may limit replaceability be restricting certain types of work to union members, replaceable tasks can be redefined as less replaceable.

Key Learning Objectives

After a thorough reading of Chapter 9, you should be able to:

1. describe how the Greeks and Romans perceived social class.

2. identify the key elements associated with Marx's concept of social class.

3. distinguish Marx's understanding of class consciousness and false consciousness.

4. specify the interaction among Weber's three dimensions of stratification

(class/property, status/prestige and party/power).

5. define and provide examples of status inconsistency.

6. distinguish achieved and ascribed status and specify how each impacts mobility.

7. indicate how achieved and ascribed status are related to class and caste.

8. clearly distinguish structural and exchange mobility.

9. discuss how cultural capital may vary by social class.

10. state why Marx believed that the creation of a "classless society" would reduce inequality.

11. critically evaluate Marx's concept of the "classless society" by alluding to the work of Dahrendorf and Mosca.

12. compare and contrast the functionalist, evolutionary and conflict approaches to stratification.

13. specify the association between a task's functional importance, its replaceability and stratification.

14. note how increasing specialization can promote stratification.

15. provide examples of how stratification can be maintained through the exploitation and manipulation of replaceability.

Chapter Outline

I. Concepts of Social Class
  A. Greek and Roman Perspectives
  B. Marx and the Economic Dimensions of Class
    1. Society is divided into two classes, owners (bourgeoisie) and workers (proletariat).
    2. The bourgeoisie own the means of production (land and equipment).
    3. Farmers and marginal economic producers, like migrant workers and criminals, are outside the two-class system.
    4. Conflict among groups is linked to the emergence of class consciousness.
    5. False consciousness exists when workers assume that workers and owners share common interests.
  C. Weber: Multiple Determinants of Stratification
    6. Stratification is linked to both the ownership and control of property (class).

7. Prestige (status) and power (party) can either increase access to property or be influenced by property.

II. Status Inconsistency, Social Mobility and the Classless Society

A. Status Inconsistency: Unequal Access to Property, Prestige and Power

1. A person may score high on one dimension (property) but low on another (prestige).
2. Lenski observed that persons present themselves to in terms of a higher status identification, but they are responded to in terms of a perceived lower status identification.
3. Gary Marx associated militancy among higher-status African Americans with status inconsistency experiences.

B. Social Mobility: Upward and Downward Movement within a Stratification System

1. Achieved statuses, like college graduate, are based on merit.
2. Ascribed statuses, like race, are fixed at birth.
3. Ascribed statuses play a major role in caste systems.
4. Structural mobility occurs when there is a change in the distribution of higher and lower status positions.
5. Exchange mobility takes place when some individuals or groups experience upward mobility only if others experience downward mobility.
6. Different social classes create and maintain distinctive cultures (cultural capital).
7. Wealthy persons tend to maintain cosmopolitan networks while working-class persons develop local networks.

C. Marx and the Classless Society: A Utopian Construct

1. The elimination of the private ownership of the means of production would create a classless society,
2. In a classless society, the state would own the means of production.
3. Dahrendorf claims that state control reflects inequality. Government officials possess more power.
4. Mosca linked stratification to political development and the maximizing of self-interest. Stratification is inevitable.

III. Three Theories of Stratification

A. Functionalism: Replaceability

1. Some jobs are judged more important than others and thus receive greater compensation.
2. The functional importance of a job is tied to replaceability.
3. Less replaceable jobs receive greater rewards.
4. Toy societies illustrate the association between replaceability and stratification.

B. Evolutionary Theory: Specialization

1. People preserve aspects of culture that are rewarding.
2. Since cultures become more complex over time, areas of specialization develop.

3. Some areas of specialization are considered more important than others. This promotes stratification.

C. Conflict Theory: Manipulating Replaceability

1. Persons in positions of power maintain their positions by exploiting others.
2. Professional groups and unions maximize their status by manipulating replaceability. Professional groups can restrict entrance into a field on the basis of credentials while unions can pressure employers to restrict certain tasks to union members.

Key Terms

Based on your reading of Chapter 9, you should be able to define and illustrate the key sociological concepts listed below. Page numbers are provided in parentheses as reference points.

Social Mobility (233)
Class (234)
Means of Production (235)
Bourgeoisie (235)
Proletariat (235)
Lumpenproletariat (235)
Class Consciousness (236)
False Consciousness (236)
Property (237)
Prestige (238)
Power (238)
Status Inconsistency (238)
Status Inconsistency Theories (238)
Achieved Status (239)
Ascribed Status (239)
Caste (240)
Structural Mobility (241)
Exchange Mobility (241)
Cultural Capital (242)
Utopian (244)
Anarchists (245)
Replaceability (248)
Toy Society (248)
Evolutionary Theory of Stratification (249)
Conflict Theory of Stratification (250)
Exploitation (250)
Professions (251)
Unions (251)

Key Research Studies

Listed below are key research studies cited in Chapter 9. Familiarize yourself with the major findings of these studies. Page references are provided in parentheses.

Marx and Engles: outlined a history of class struggle between owners and workers in *The Communist Manifesto* (235).

Lenski: conducted extensive studies on status inconsistency. Persons tend to present themselves in terms of their highest perceived status, but are responded to on the basis of their lowest perceived status (238-239).

Marx, Gary: linked militancy among high status African Americans to status inconsistency experiences (239).

Bourdieu: social classes maintain distinct class cultures. Upper-class cultures stress abstract thought, the arts and "intellectual" leisure activities while the lower-class cultures stress concrete experience and meeting basic needs (242).

Dahrendorf: offered a critique of Marx's concept of the classless society. Argued that the state control of the means of production does not eliminate inequality. The state is run by political specialists who have power (245-246).

Mosca: maintained that stratification is inevitable. Linked stratification to the development of political organizations and the maximization of political leaders' self-interests (246).

Davis and Moore: developed a functionalist theory of stratification . Stratification is linked to the functional importance of a task, replaceability and a system of unequal rewards (246-248).

Info Trac Search Words

Enter these search terms to conduct more extensive investigations of key topics introduced in Chapter 9.

Social Inequality

Prestige and Power

Class and Caste

Choose one of the studies listed. Note the interaction of class and caste in the study selected. Identify the issue or issues being impacted by class and/or caste.

Cultural Capital

Select one of the studies cited. How is cultural capital being defined in your selected study?

Social Stratification

Subdivisions: Research

Multiple Choice

Answers and page references are provided at the chapter end.

1. The Romans used the term *classis* to divide the population into groups in order to
   a. be able to collect taxes.
   b. separate different religious groups.
   c. determine who was fit for military service.
   d. identify potential political threats.
   e. reduce conflict.

2. Marx maintained that machinery, land and investment capital are examples of
   a. false consciousness.
   b. the bourgeoisie.
   c. caste systems.
   d. the lumpenproletariat.
   e. the means of production.

3. The owners of the means of production are members of the
   a. lumpenproletariat.
   b. middle class.
   c. bourgeoisie.
   d. proletariate.
   e. false consciousness class.

4. The lumpenproletariat includes
   a. vagrants.
   b. migrant workers.
   c. beggars.
   d. criminals.
   e. all of the above

5. Similar placement, the sharing of mutual interests and the sharing of a common enemy are all characteristics of
   a. social mobility.
   b. status inconsistency.
   c. replaceability.
   d. class consciousness.
   e. status and party.

6. According to Marx, the single most important factor determining rank or social position is
   a. gender.
   b. property ownership.
   c. religious background.
   d. age.
   e. race and ethnicity.

7. Weber identified __________ as the three most important factors influencing stratification.
   a. property, prestige and power
   b. age, gender and ethnicity
   c. power, religion and politics
   d. race, prestige and age
   e. politics, occupation and power

8. Social honor or respect is an example of this dimension of stratification.
   a. bourgeoisie
   b. proletariat
   c. property
   d. prestige
   e. power

9. In many societies the police are granted the authority to enforce the laws. This is an example of
   a. property.
   b. prestige.
   c. power.
   d. exchange mobility.
   e. structural mobility.

10. A person may have won the lottery but still be denied country club membership. This is an example of
   a. false consciousness.
   b. replaceability.
   c. status inconsistency.
   d. exploitation.
   e. the lumpenproletariat.

11. Lenski argued that when persons interact with others, they tend to
   a. emphasize their lowest claim to rank and deemphasize their highest claim.
   b. emphasize their highest claim to rank and deemphasize their lowest claim.
   c. emphasize neither their highest nor the lowest claim to rank
   d. emphasize both their highest and lowest claim to rank.
   e. deemphasize both their highest and lowest claim to rank.

12. According to __________, more affluent members of ethnic minority groups would tend to support more liberal political parties.
   a. status inconsistency theories
   b. exchange theories
   c. the functionalist theory of stratification.
   d. conflict theories of stratification.
   e. social evolutionary theories of stratification

13. Race, gender and place of birth are examples of
   a. structural statuses.
   b. exchange statuses.
   c. inconsistent statuses.
   d. achieved statuses.
   e. ascribed statuses.

14. Caste systems are characterized primarily by
   a. a very high degree of social mobility.
   b. no social mobility.
   c. powerful ascriptive statuses.
   d. a and c
   e. b and c

15. The major determinant of status in the U.S., Canada and other highly industrialized countries is
   a. ascriptive status.
   b. achievement.
   c. race.
   d. gender.
   e. age.

16. If some members of a society are able to experience upward mobility only if other members experience downward mobility, then this society would be experiencing
   a. exchange mobility.
   b. structural mobility.
   c. ascriptive mobility.
   d. status mobility.
   e. false mobility.

17. Speech styles, dress, body language, interests and tastes can be examples of
   a. the means of production.
   b. false consciousness.
   c. the lumpenproletariat.
   d. cultural capital.
   e. replaceability.

18. According to Bourdieu's research on culture and social class, upper social classes stress each of the following except
   a. meeting basic needs.
   b. abstract thought.
   c. "intellectual" leisure activities.
   d. involvement in the arts.
   e. appreciation for great literature.

19. Working-class individuals tend to develop extensive __________ networks.
   a. cosmopolitan
   b. structural
   c. exchange
   d. local
   e. utopian

20. The concept of the classless society was developed by
   a. Weber.
   b. Lenski.
   c. Marx.
   d. Mosca.
   e. Dahrendorf.

21. The classless society is an example of a/an
   a. anarchy.
   b. monopoly.
   c. dictatorship.
   d. highly stratified society.
   e. utopian society.

22. Marx argued that in the classless society everyone becomes a
   a. lumpenproletarian.
   b. farmer.
   c. bourgeoisie.
   d. proletarian.
   e. none of the above

23. A classless society would be created according to Marx, if __________ were to gain control of the means of production.
   a. religious leaders
   b. the state
   c. military leaders
   d. business owners
   e. all of the above

24. Which of the following statements about Weber's and Dahrendorf's critique of Marx's classless society is false?
   a. Marx did not say that the classless society would not be stratified.
   b. The state is controlled by political specialists who possess more power.
   c. In the classless society, the capitalist boss is replaced by a communist boss.
   d. Significant variations in the control of the means of production will exist even if the private ownership of the means of production is outlawed.
   e. In the classless society, the communist boss is replaced by a capitalist boss.

25. Mosca associates increasing stratification with the development of
   a. religious institutions.
   b. a strong military.
   c. political organizations.
   d. strong communist establishments.
   e. weak family structures.

26. Functional importance of jobs, a system of unequal rewards and replaceability are key concepts associated primarily with the __________ theory of stratification.
    a. conflict.
    b. functionalist
    c. exchange
    d. opportunity
    e. social evolutionary

27. Data from the *International Social Survey Program* suggest that the majority of respondents in industrial societies agree that people should not be expected to sacrifice years of study to become a lawyer or doctor unless they expect to receive significant financial rewards in return. These data provide support for the claims of the __________ theory of stratification.
    a. functionalist
    b. social evolutionary
    c. opportunity
    d. exchange
    e. conflict

28. In the toy society "ay" is the least replaceable person because "ay" is the only survivor who can produce
    a. heat.
    b. heat and food.
    c. heat, food and water.
    d. heat, food, water and air.
    e. none of the above

29. Many school districts are providing more funding for science and math and less funding for music and the arts. This creates inequality and stratification since all areas of specialization are not weighted equally. This example best illustrates the __________ theory of stratification.
    a. conflict
    b. social evolutionary
    c. exchange
    d. functionalist
    e. opportunity

30. Exploitation and manipulation of replaceability are concepts most often associated with the __________ theory of stratification.
    a. opportunity
    b. exchange
    c. conflict
    d. social evolutionary
    e. functionalist

True/False

Answers and page references are provided at the chapter end.

1. The upward and downward movement of individuals or groups within a stratification system is known as social mobility.

2. Owners are to workers as proletariat is to bourgeoisie.

3. According to Weber, the three major determinants of stratification are race, class and gender.

4. In his study of the association between militancy and experiences of status inconsistency among African Americans, Gary Marx observed that higher-status African Americans were more militant about changing racial conditions than were lower-status African Americans.

5. A college graduate would be an example of an ascribed status.

6. Ascribed statuses play the major role in defining status and position in caste systems.

7. Structural mobility is experienced when there is no change in the distribution of statuses within a society and some individuals or groups move up only if other individuals or groups move down.

8. More affluent members of society tend to develop cosmopolitan networks; whereas, working-class individuals are more likely to form extensive local networks.

9. According to the functionalist theory of stratification, replaceability increases as the functional importance of a task increases.

10. The evolutionary of stratification associates increasing stratification with the accumulation of culture and increasing specialization.

Short Answer Questions

These short answer questions are provided to test your knowledge and understanding of the basic sociological concepts presented in Chapter 9. Page references for answers are included at the chapter end.

1. How did the Greeks and Romans perceive social class?

2. What is the nature of the association among the bourgeoisie, the proletariat and the ownership of the means of production?

3. Identify Weber's three dimensions of stratification.

4. Define status inconsistency and provide a concrete example.

5. Provide an example of an achieved status and an ascribed status.

6. How do structural and exchange mobility differ?

7. What is the nature of the association between social class and cultural capital?

8. According to Mosca, why is stratification inevitable?

9. How does the toy society illustrate the association between replaceability and stratification?

10. Provide examples of how professional groups and labor unions are able to manipulate replaceability.

Essay Questions

These questions are designed to test your understanding of key sociological concepts presented in Chapter 9 and your ability to apply these insights to concrete situations.

1. With the decline of jobs in the manufacturing sector due to automation, many fear that the U.S. middle class is shrinking. To what extent is the U.S. becoming a two-class society?

2. What are the major determinants of social position in U.S. society? Are these determinants primarily based on ascribed or achieved status?

3. Is Affirmative Action primarily an example of structural or exchange mobility? Justify your answer.

4. Funding for various educational programs within a given school district may vary significantly. How could the social evolutionary theory of stratification provide a plausible explanation of the program funding differences?

5. Identify the major strengths and weaknesses of the conflict theory of stratification.

Answers

Multiple Choice

1. a (234)
2. e (235)
3. c (235)
4. e (235)
5. d (235-236)
6. b (236-237)
7. a (237-238)
8. d (238)
9. c (238)
10. c (238-239)
11. b (239)
12. a (239)
13. e (239)
14. e (240)
15. b (240)
16. a (241-242)
17. d (242)
18. a (242)
19. d (243)
20. c (243)
21. e (243-244)
22. d (245)
23. b (245)
24. e (245-246)
25. c (246)
26. b (246-248)
27. a (247)
28. d (248-249)
29. b (249-250)
30. c (250-251)

True/False

1. T (233)
2. F (235)
3. F (237)
4. T (239)
5. F (239)
6. T (240)
7. F (241-242)
8. T (243)
9. F (246-248)
10. T (249-250)

Short Answer

1. (234)
2. (235)
3. (237-238)
4. (238-239)
5. (239)
6. (241-242)
7. (242-243)
8. (246)
9. (248-249)
10. (251)

CHAPTER 10

COMPARING SYSTEMS OF STRATIFICATION

Extended Abstract

The degree of stratification present within a society varies by the type of society. In this chapter incidence of stratification in hunting and gathering, horticultural, pastoral, agrarian and industrial societies is assessed. Significant attention is devoted also to social mobility and status attainment in industrial societies.

Hunting and gathering societies are small, highly mobile and less stratified. As societies become more productive and develop stronger ties to one location, as is the case with horticultural, pastoral and agrarian societies, stratification increases. Data from the *Standard Cross-Cultural Sample* indicate that stratification is lowest among nomadic societies and highest among societies that establish more permanent residence.

The increasing occupational specialization and urbanization associated with agrarian societies is linked to their ability to produce surplus food. Data from the *Standard Cross-Cultural Sample* indicate that warfare is more common among societies experiencing higher levels of agricultural development and stratification increases as agricultural productivity increases. In agrarian societies, the affluent tend to control the military and possess a distinct culture. These cultural differences may be used to maintain and reinforce high levels of stratification. Consequently, social scientists generally agree that stratification is most pronounced in agrarian societies. In a brief excursus, Stark outlines how the European feudal system developed and how agrarian elites were able to control military protection and guarantee resources for knights. Military specialists or knights were granted the authority to levy taxes, and exploit the peasants in exchange for military protection.

Since data presented for agrarian societies indicate that stratification increases as productivity increases, one might assume that stratification would be higher in industrial societies as industrialization enhances productivity. However, data from the *Nations of the Globe* data source indicate that stratification decreases as industrial development and modernization increase. Stark attributes this reversal in the productivity - stratification association to two important factors. First, workers in industrial societies must acquire more skills. This makes them less replaceable. Second, skilled workers are more powerful and can bargain for better wages. These factors tend to reduce the degree of stratification present within industrial societies, and ascriptive sources of status are increasingly replaced with those based on performance.

In the second half of this chapter, Stark reviews numerous studies of social mobility and status attainment in highly industrialized societies. Lipset and Bendix conducted a classic study on comparative social mobility in industrialized nations. They observed that the degree of total mobility achieved by these nations was roughly equivalent and that most industrial societies experience a greater degree of upward mobility. This finding refuted the general belief that total mobility is unusually high in the U.S. Subsequent comparative studies on social mobility in industrialized societies have indicated that the U.S. is characterized by a high degree of long-distance mobility. Individuals can experience significant increases or decreases in status during their lifetime. Data from the *International Social Survey Program* indicate that U.S. residents do not perceive income differences as being too large and attribute economic success to hard work rather than wealthy family connections or important political connections. Status attainment models attempt to measure how status is attained and passed from one generation to another. Blair and Duncan conducted a classic study in status attainment. These researchers concluded that the status attainment is primarily influenced by education. Subsequent research by Cohen and Tyree confirm this finding and suggest that marital status is also a major determinant of family income. Research by Mare and Tzeng indicates that having older parents may convey certain economic advantages. Older parents tend to have higher incomes, higher levels of education and smaller families. Smaller families mean that more economic resources can be devoted to rearing each child. In a classic study on status attainment in Canada, Porter hypothesized that the prospects for status attainment in Canada would be lower than those in the U.S. since Canadian opportunities for advancement are more restricted and since stratification varies by ethnicity and immigrant status. Findings by Porter and other colleagues have failed to substantiate these claims. However, research has shown that female Canadian workers tend to come from higher-status backgrounds and work in higher-status occupations. Studying status attainment patterns in the U.S., Hout notes that there has been a decline in structural mobility and an increase in exchange mobility. Furthermore, college education exerts a stronger impact on status attainment than does family background.

The two remaining studies focus on network structures and mobility. The first study by Erickson looks at differences in cultural variety by social class among Canadians. Erickson concludes that upper-class cultural networks are more diverse reflecting a greater awareness and acquisition of "highbrow" culture and "mass" culture. The upper-class cultural advantage is linked to exposure to more cultural variety rather than differences in culture. In analyzing factors associated with Chinese job acquisition strategies, Bian observed that jobs are acquired primarily through strong local networks. Family and friends play a big role in helping persons obtain a job. The opposite pattern is generally observed in the U.S. Here acquaintances (cosmopolitan networks) rather than family and friends play a greater role in helping persons acquire a job.

Key Learning Objectives

After a thorough reading of chapter 10, you should be able to:

1. state why stratification is lower in hunting and gathering societies.
2. identify the nature of the association between fixity of residence and stratification.
3. describe the distinctive structural features of agrarian societies.
4. identify the nature of the association between warfare and agricultural development.
5. describe the association between stratification and agricultural productivity.
6. state why social scientists claim that stratification is most pronounced in agrarian societies.
7. describe the major characteristics of feudalism.

8. critically assess the productivity - stratification association in industrial societies.

9. identify major patterns of total mobility, upward mobility and long-distance mobility among highly industrialized societies.

10. specify the association between hard work and status attainment in the U.S.

11. demonstrate how education, marital status and parent's age impact status attainment.

12. state the major findings from the studies on status attainment in Canada by Porter and other colleagues.

13. identify recent trends in structural and exchange mobility in the U.S.

14. critically evaluate class differences in the acquisition of extensive cultural networks.

15. specify how local and cosmopolitan networks impact job acquisition in China and the U.S.

Chapter Outline

I. Stratification in Simple Societies
   A. Hunting and Gathering Societies: Small, High Mobility and Less Stratified
   B. Herding Societies and Horticultural Societies: More Complex, Productive and Stratified
   C. Stratification Increases as Permanence of Residence Increases

II. Stratification in Agrarian Societies
   A. Agrarian Societies and Surplus Food Production
   B. Surplus Food Production, Occupational Specialization and Urbanization
   C. Warfare Increases as Agricultural Development Increases
   D. Stratification Increases as Agricultural Productivity Increases
   E. Military Domination and Agrarian Societies
      1. The elite dominate military power in agrarian societies.
      2. In European feudal societies, knights were able to exploit peasants through excessive taxation in exchange for military protection.
   F. Ruling Elites, Cultural Differences and Stratification
   G. Agrarian Societies: Most Highly Stratified

III. Stratification in Industrial Societies
   A. Stratification Decreases as Productivity Increases
      1. Workers Possess more skills and are less replaceable.
      2. Skilled workers are more powerful and can bargain for better wages.
   B. Social Mobility and Achieved Status

IV. Social Mobility in Highly Industrialized Societies

A. Lipset and Bendix: Upward and Total Mobility among Industrial Nations Is Similar
B. Long - Distance Mobility: A Distinctive U.S. Feature
C. Hard Work: An Important Determinant of Success in the U.S.

V. Status Attainment Models
   A. Education Is The Key: Blau and Duncan
   B. Dissimilar Status Attainment among Siblings: Jencks
   C. Two-Earner Families and Status Attainment: Cohen and Tyree
   D. Older Parents and Status Attainment: Mare and Tzeng
      1. Family income is higher among older parents.
      2. Older parents tend to have more education.
      3. Older parents maintain smaller families and can provide more economic resources per child.
   E. Status Attainment in Canada: Porter and Colleagues
      1. Porter hypothesized that status attainment is lower in Canada than in the U.S. due to fewer occupational opportunities and to stratification by ethnic an immigrant status.
      2. Data reveal that status attainment is similar for all groups.
      3. Female workers tend to come from higher-status backgrounds and are in higher-status occupations.
   F. Changing U.S. Mobility Patterns: Hout
      1. Structural mobility is decreasing, but exchange mobility is increasing.
      2. A college education enhances social mobility.

VI. Networks, Social Class and Economic Success
   A. Cultural Networks and Social Class in Canada: Erickson
      1. Culture is a reflection of social network diversity
      2. Upper classes maintain more culturally diverse networks
   B. Social Networks and Economic Success: Bian
      1. In the U.S. weak, cosmopolitan networks strongly influence job acquisition.
      2. Strong, local networks have a greater impact on job acquisition in China.

Key Terms

Based on your reading of Chapter 10, you should be able to define and illustrate the key sociological concepts listed below. Page numbers are provided in parentheses as reference points.

Hunting and Gathering Societies (258)
Agrarian Societies (261)
Feudalism (265)
Industrial Societies (268)
Industrialization (268)
Long-Distance Mobility (272)
Status Attainment Model (274)
Vertical Mosaic (276)
Highbrow Culture (278)
Network Variety (279)

Key Research Studies

Listed below are key research studies cited in Chapter 10. Familiarize yourself with the major findings of these studies. Page references are provided in parentheses.

Sorokin: studied the agrarian period of eleven industrial European nations and concluded that they were at war 46 percent of the time (262).

Tocqeville: observed wealth and social mobility in the U.S. in the 1830s and assumed that upward mobility was common (270).

Lipset and Bendix: studied social mobility in highly industrial societies and noted little variation in total mobility. The high levels of social mobility characteristic of the U.S. is not distinctive (271-272).

Blau and Duncan: long-distance mobility is a characteristic feature of U.S. mobility, and education significantly impacts status attainment (273-274).

Jencks: status attainment differences among brothers are significant (274).

Cohen and Tyree: education and marital status exert a strong influence on status attainment especially among the poor (274).

Mare and Tzeng: having older parents provides more economic advantages for children. Older parents have higher incomes, are more educated and produce smaller families (275).

Porter and colleagues: discovered that status attainment in Canada and the U.S. is the same. Hypothesized differences in the status attainment of Canadians of different ethnic backgrounds and immigrant status were not supported by the data. Canadian working women tend to come from higher-status backgrounds and are in higher-status occupations (275-277).

Hout: structural mobility is decreasing in the U.S., but exchange mobility is increasing. College education is identified as an important factor impacting status attainment (277-278).

Erickson: culture is a product of social network diversity rather than social class differences. Upper-class cultural networks are more extensive and diverse (278-280).

Bian: strong, local networks play a major role in job acquisition in China; whereas, weak, cosmopolitan networks play a stronger role in job acquisition in the U.S. (280).

Info Trac Search Words

Enter these search terms to conduct more extensive investigations of key topics introduced in Chapter 10.

Hunting and Gathering Societies

Select one of the articles listed. What aspects of the hunter-gatherer life style are addressed in the selected article?

Comparative Sociology

Industrial Societies

Protestant Work Ethic

Status Attainment

Choose one of the studies listed. How is status attainment conceptualized? Note relevant factors influencing status attainment identified in the study.

Multiple Choice

Answers and page references are provided at the chapter end.

1, According to Stark, simple societies include
  a. agrarian, industrial and information societies.
  b. herding, agrarian and industrial societies.
  c. horticultural, herding and industrial societies.
  d. industrial, hunting and gathering and agrarian societies.
  e. hunting and gathering, horticultural and herding societies.

2. Stratification appears to be lowest in __________ societies.
  a. hunting and gathering
  b. herding
  c. horticultural
  d. agrarian
  e. industrial

3. Which one of the following statements about hunter-gatherers is false?
  a. These societies are small and often number approximately fifty persons.
  b. Hunter-gatherers are nomadic.
  c. Compared to other type societies, hunter-gathers are less stratified.
  d. Hunter-gathers have been known to be very deadly hunters.
  e. Hunter-gatherers are not known to be proficient in preserving food.

4. Stratification generally increases in
   a. smaller societies.
   b. societies characterized by universal poverty.
   c. less secure and stable societies.
   d. more productive societies.
   e. none of the above

5. Data from the *Standard Cross-Cultural Sample* indicate that the degree of stratification is lowest
   a. among sedentary populations.
   b. among nomadic populations.
   c. among permanent populations.
   d. in agrarian societies.
   e. in industrial societies.

6. The use of plows and domesticated animals would be greatest in
   a. hunting and gathering societies.
   b. herding societies.
   c. horticultural societies.
   d. agrarian societies.
   e. pastoral societies.

7. Although increasing specialization of labor is present in agrarian societies, Sjoberg estimates that in the typical agrarian society, as much as __________ of the population would be actively involved in farming.
   a. 5%
   b. 10%
   c. 25%
   d. 50%
   e. 90%

8. Data from the *Standard Cross-Cultural Sample* indicate that frequency of external war is
   a. higher among societies characterized by a low level of agricultural development.
   b. higher among societies characterized by a medium level of agricultural development.
   c. higher among societies characterized by a high level of agricultural development.
   d. all off the above
   e. none of the above

9. Which of the following statements is true?
   a. Stratification increases as level of agricultural productivity increases.
   b. Stratification increases as level of agricultural productivity decreases.
   c. Stratification decreases as level of agricultural productivity decreases.
   d. a and c are true
   e. b and c are true

10. Surplus production has been associated with
    a. increasing job specialization.
    b. increasing urbanization.
    c. the emergence of slavery.
    d. political development.
    e. all of the above

11. In feudal societies
    a. land ownership was based on religious obligations.
    b. peasants had more power than knights.
    c. taxation was linked to protection.
    d. knighthood was a ceremonial title.
    e. knights greatly outnumbered peasants.

12. Sociologists generally agree that stratification is greatest in __________ societies.
    a. agrarian
    b. hunting and gathering
    c. industrial
    d. horticultural
    e. herding

13. With increasing industrialization
    a. the middle class began to expand.
    b. the gap in standard of living between the rich and the poor increased.
    c. unskilled jobs proliferated.
    d. welfare programs were abolished.
    e. status was based more on ascription than achievement.

14. According to the *Nations of the Globe* data source, the percent of total national income going to the richest 10 percent of families decreases as level of economic development increases. This finding illustrates
    a. a negative correlation and an increase in stratification.
    b. a negative correlation and a decrease in stratification.
    c. a positive correlation and an increase in stratification.
    d. a positive correlation and a decrease in stratification.
    e. no correlation and no change in stratification.

15. Some researchers maintain that stratification is lower in industrial societies because
    a. laborers work harder rather than smarter.
    b. slavery is introduced as a more efficient means of controlling labor costs.
    c. workers are less educated and more replaceable.
    d. more services can be provided through taxation.
    e. workers are more educated and less replaceable.

16. In *Democracy in America*, Alexis de Tocqueville noted that many Americans
    a. were upwardly mobile and proud of their humble beginnings.
    b. were downwardly mobile and ashamed of their humble beginnings.
    c. rarely achieved great wealth or much education.
    d. still in what was essentially a feudal society.
    e. were facing unemployment as a result of increasing industrialization.

17. Lipset and Bendix observed that compared to other industrialized nations
    a. more upward mobility is experienced in the U.S.
    b. more total mobility is experienced in the U.S.
    c. the U.S. is not distinctive with respect to total or upward mobility.
    d. more people in the U.S. experienced downward mobility than upward mobility.
    e. none of the above

18. The daughter of a textile worker becomes a lawyer for the textile labor union. This is an example of
    a. structural mobility.
    b. exchange mobility.
    c. family connections.
    d. long-distance mobility.
    e. downward mobility.

19. U.S. residents overwhelmingly insist that "getting ahead in life" is due to
    a. hard work.
    b. family connections.
    c. Affirmative Action.
    d. winning the lottery.
    e. political connections.

20. According to Blau and Duncan the major determinant of status attainment is
    a. age.
    b. gender.
    c. education.
    d. race.
    e. occupation.

21. Two important determinants of status attainment according to Cohen and Tyree are
    a. race and ethnicity.
    b. coming from a wealthy family and political connections.
    c. education and marital status.
    d. hard work and luck.
    e. age and highbrow culture.

22. Starting a family when the parents are older may be advantageous for children because
    a. family income tends to decrease over time.
    b. persons who delay having children generally have higher educational attainment levels.
    c. they produce larger families.
    d. they are able to provide better supervision since they spend less time with their children.
    e. fewer financial resources will be available for each child and they will learn to share.

23. The vertical mosaic is a term coined by Porter and refers to stratification based on
    a. age and gender.
    b. property, power and prestige.
    c. control of the means of production.
    d. ascriptive status.
    e. ethnicity.

24. The study of status attainment by Porter and others indicates that
    a. status attainment opportunities are greater in the U.S. than in Canada.
    b. status attainment opportunities are greater in Canada than in the U.S.
    c. status attainment opportunities in Canada and the U.S. are essentially the same.
    d. Canadian women tend to come from lower-status family backgrounds.
    e. Canadian working women tend to have lower-status jobs compared to men.

25. The correlation between son's education and son's occupational prestige is 0.62 for English-speaking Canadians and 0.61 for French-speaking Canadians. This means that
    a. ethnicity is a major determinant of status attainment.
    b. ethnicity does not appear to influence status attainment.
    c. French-speaking Canadians have more status attainment opportunities than English-speaking Canadians.
    d. native- born Canadians have more status attainment opportunities than foreign-born Canadians.
    e. Canadian fathers have experienced more opportunities for status attainment than have their sons.

26. Hout has observed that recently in the U.S. there has been
    a. a decrease in structural mobility and an increase in exchange mobility.
    b. an increase in structural mobility and a decrease in exchange mobility.
    c. an increase in structural and exchange mobility.
    d. a decrease in structural and exchange mobility.
    e. no change in either structural or exchange mobility.

27. Hout argues that the U.S. stratification system is becoming more open because
   a. race and ethnicity are no longer major determinants of status.
   b. more people are able to acquire full-time jobs with good benefit packages.
   c. more people are choosing to remain single.
   d. a larger proportion of the labor force now has college degrees.
   e. gender differences in income have been largely eliminated.

28. An appreciation for art, opera, Shakespeare and gourmet food are traditional examples of __________ culture.
   a. pop
   b. mass
   c. highbrow
   d. feudal
   e. lowbrow

29. In her study assessing differences in cultural acquisition by social class, Erickson observed that persons from upper-class backgrounds possessed more knowledge and cultural awareness in the areas of
   a. sports.
   b. art.
   c. books.
   d. magazine stories.
   e. all of the above

30. In China job acquisition is significantly influenced by the presence of
   a. local networks characterized by strong social ties.
   b. local networks characterized by weak social ties.
   c. cosmopolitan networks characterized by weak social ties.
   d. cosmopolitan networks characterized by strong social ties.
   e. a high degree of network variety.

True/False

Answers and page references are provided at the chapter end.

1. Hunting and gathering societies are larger in size, tend to establish permanent residence and are highly stratified.

2. Increasing occupational specialization and urbanization are an outcome of surplus food production in simple societies.

3. Stratification tends to increase as level of agricultural productivity increases.

4. Under the feudal system, peasants were taxed heavily in order to pay for the military protection provided by knights.

5. Ruling elites in agrarian societies were often from the same ethnic background as their subjects.

6. In industrial societies, status is based more often on achievement than on ascription.

7. In industrial societies, workers quickly realized that they could experience more economic success if they worked harder rather than working smarter.

8. Compared to European industrialized societies, the U.S. appears to be characterized by a higher degree of long-distance mobility.

9. Jencks has observed that status attainment levels of brothers is essentially the same.

10. Contrary to what Porter hypothesized, ethnicity exerts little impact on status attainment in Canada.

Short Answer Questions

These short answer questions are provided to test your knowledge and understanding of the basic sociological concepts presented in Chapter 10. Page references for answers are included at the chapter end.

1. How is stratification determined in hunting and gathering societies?

2. What is the association among increased productivity, increased ties to the land, more complex social development and stratification?

3. Assess the impact of increased agricultural development on frequency of warfare and of level of agricultural productivity on degree of stratification.

4. How was stratification maintained under the feudal system?

5. Identify two social changes associated with industrialization that led to a reduction in stratification in industrial societies.

6. Does the U.S. differ significantly from other industrial nations with respect to degree of total mobility and upward mobility?

7. According to Blau and Duncan, what variable has the greatest impact on status attainment?

8. What economic advantages accrue to a family when parents begin their family at an older age?

9. What is the vertical mosaic, and does status attainment in Canada and the U.S. vary significantly?

10. How do the U.S. and China differ with respect to the way that social networks are linked to economic success?

Essay Questions

These questions are designed to test your understanding of key sociological concepts presented in Chapter 10 and your ability to apply these insights to concrete situations.

1. Present evidence that demonstrates why stratification would be higher in herding and horticultural (gardening) societies than in hunting and gathering societies.

2. Many sociologists claim that stratification is most pronounced in traditional agrarian societies. Would the mechanization of agriculture increase or decrease stratification in agrarian societies?

3. Stratification is reduced in industrial societies due to the presence of a skilled labor force that is less replaceable. With the increasing use of automation in manufacturing and other areas of work, how will is this affect worker replaceability? Will increased productivity now lead to increased stratification?

4. Given the growth of the Hispanic American and Asian American population in the U.S., how would Porter's concept of the vertical mosaic apply to the analysis of stratification, social mobility and status attainment in the U.S.?

5. Erickson maintains that cultural inequality is more a reflection of a hierarchy of knowledge rather than a hierarchy of tastes. How would a researcher go about evaluating this claim?

Answers

Multiple Choice

1. e (258)
2. a (258)
3. d (258-259)
4. d (260)
5. b (260)
6. d (261)
7. e (261)
8. c (262-263)
9. d (262-263)
10. e (261- 263)
11. c (265)
12. a (267-268)
13. a (268)
14. b (268)
15. a (269)
16. d (270)
17. c (271-272)
18. d (272)
19. a (273)
20. c (274)
21. c (274)
22. b (275)
23. e (276)
24. c (275-277)
25. b (276)
26. a (278)
27. d (278)
28. c (278)
29. e (279)
30. a (280)

True/False

1. F (258-260)
2. F (261)
3. T (263)
4. T (265)
5. F (263, 266)
6. T (268)
7. F (268)
8. T (272)
9. F (274)
10. T (275-276)

Short Answers

1. (260-261)
2. (260-263)
3. (262-263)
4. (265)
5. (268-270)
6. (271-272)
7. (274-275)
8. (275)
9. (276-277)
10. (280)

CHAPTER 11

RACIAL AND ETHNIC INEQUALITY AND CONFLICT

Extended Abstract

Intergroup conflict is grounded in racial and ethnic group inequality. Sociologists tend to define race in terms of society's response to perceived physical differences among group members; whereas, ethnicity is conceptualized as society's response to a group's perceived cultural differences. Societal responses leading to unequal treatment may be expressed in terms of hatred, prejudice or discrimination. Outcomes of intergroup conflict include assimilation, accommodation and cultural pluralism, extermination, expulsion and segregation. Racial and ethnic groups may be expected to conform, cultural distinctiveness may be tolerated or groups maybe eliminated, driven away or separated.

Stark next turns to a more focused discussion of prejudice. Several important theories of prejudice are reviewed, and the link between status inequality and prejudice and between identifiability and prejudice is specified. A classic theory of prejudice is Adorno's theory of the authoritarian personality. According to this theory, persons are only able to accept the norms and practices of their own group. Persons who are different threaten them. Prejudice becomes a defense mechanism that reduces the anxiety produced by those who are different.
Other theories of prejudice link prejudice to low self-esteem, low levels of educational attainment and low income. Another classic theory of prejudice is Allport's theory of contact. Here prejudice is linked to the nature of intergroup contact. Allport argued that contact among groups competing for available opportunities stimulates prejudice while cooperation among groups pursuing common goals reduces prejudice. Studying summer camp experiences of young boys, the Sherifs observed that competitive athletic activities generated the development of hostile stereotypes that were applied to friends who were members of different teams.

The studies by Allport and the Sherifs suggest that status inequality may lead to prejudice. It is within this context that Stark looks at slavery as an institutionalized mechanism of status inequality that stimulated prejudice. Status inequality was epitomized in the master-slave relationship, and in a well-known study on U.S. racial inequality Myrdal maintained that the paradox of American democratic ideals and racist practices, the "American dilemma," has characterized U.S. society since the country's inception. Many sociologists argue that reduction in status inequality will lead to reductions in intergroup prejudice and discrimination. The attainment of economic parity between Catholics and Protestants and the decline in prejudice and discrimination directed toward Catholics is cited as support for the status inequality hypothesis.

Stark notes that racial and ethnic minority groups may be perceived as an economic threat by dominant groups because they are willing to work for less. This may promote racial and ethnic hostility among dominant group members. Bonacich argues that racial and ethnic groups may work for lower wages because low wages could represent an increase in standard of living, persons may not be aware that they are being exploited, persons may lack political power and persons may be motivated by different economic concerns. To restrict the impact of a low-wage labor source on dominant group wages, dominant groups have placed restrictions on immigration, tried to encourage minority group participation in lower-status occupations and have allowed minority groups to become scapegoats (middleman minorities) for dominant group class conflicts.

Identifiability may also increase prejudice and discrimination. Within a status inequality context, conflict between clearly identifiable racial and ethnic groups is harsher as entire groups are stereotyped and stigmatized. During the Great Depression, migrants from Oklahoma to California represented a low-wage labor source that was clearly identifiable. The rural life style and the Southern dialect of the "Okies" made them easier targets of prejudice and discrimination. On the other hand, prejudice and discrimination should decline among identifiable minority groups that attain status equality with dominant groups. This claim is addressed by briefly considering the changes in the socioeconomic experiences of Japanese Americans and Japanese Canadians.

Japanese American experiences of prejudice and discrimination are well known. Laws restricting land ownership were passed in the early twentieth century, immigration from Japan was restricted in 1924 and during WWII, a significant number of Japanese Americans were placed in internment camps. Nevertheless, Japanese Americans have become very successful. Their success appears to be linked to educational attainment and entrepreneurial activity. As early as 1930, Japanese Americans of college age were twice as likely as native-born whites to be enrolled in school. Japanese Americans also experienced economic success prior to WWII running their own farms, establishing small businesses and maintaining credit unions. Following WWII, the demand for highly educated people was great, and Japanese Americans were able to benefit from an increase in high status occupational opportunities. By 1970 Japanese American and white earning ratios were equivalent. The high rate of intermarriage among Japanese Americans further reflects a reduction in Japanese American experiences of prejudice and discrimination. Stark notes that the experiences of Japanese Canadians mirror that of Japanese Americans.

Stark next introduces three mechanisms that have been associated with racial and ethnic group mobility in North America. These mechanisms are geographic concentration, internal economic development and the development of a middle class. Although racial and ethnic minority groups may be geographically concentrated as a result of discrimination, geographic concentration enables groups to enhance their economic and political resources.

Hispanic Americans are the largest U.S. minority group. They represent a diverse population that is geographically concentrated. Mexican Americans reside primarily in the Southwest. Puerto Rican Americans are concentrated in New York, and many Cuban Americans reside in Florida. However, have Hispanic Americans attained status equality in the U.S., and does status equality exist within the Hispanic American community? Available data on the economic experiences of Hispanic Americans indicate that compared to the total U.S. population, Hispanic American families are more than twice as likely to live below the poverty level. However, within the Hispanic American community, Puerto Rican American families are almost twice as likely as Cuban American families to live below the poverty level. Also, among young adults aged 25-35, Hispanic Americans are only half as likely as the total population to be college graduates, but Cuban Americans have reached parity. These data suggest that the Hispanic American community has not attained status equity with the total population, but less status inequality has been experienced by Cuban Americans. If reductions in status inequality lead to reductions in prejudice and discrimination, Cuban American should report fewer experiences of discrimination. Data from a national survey of Hispanic Americans confirm this trend. Fewer Cuban Americans report that they have been discriminated against.

African Americans represent the second largest U.S. minority group, and approximately half of all African Americans still live in the South. In recent decades African Americans have experienced some significant gains in education and income. Since 1960, the African American-white gap in high school and college educational attainment has closed, but African Americans are only two-thirds as likely as whites to have completed college. Likewise, from 1967 to 2000 the African American-white earnings ratios for working couples increased from .71 to .89. However, when all households are compared, the earnings ratio stands at .58. Reductions in status inequality are evident, but status inequality still persists.

On the other hand, survey data indicate that as an increasing number of U.S. residents live in integrated neighborhoods, interracial friendships are more common. Interracial marriages are increasing also. Firebaugh and Davis' analysis of 1972-1984 *General Social Survey* data indicates that prejudice against African Americans declined steadily throughout this period and the decline was most evident in the South. Finally, Stark argues that the legacy of slavery, the lack of a homeland, identifiability and the size of the African American community have had a long-term, collective impact on African American experiences of prejudice, discrimination and mobility.

Key Learning Objectives

After a thorough reading of Chapter 11, you should be able to:

1. portray the interaction among inequality, intergroup conflict, prejudice and discrimination.
2. distinguish race and ethnicity.
3. identify major outcomes of integroup conflict such as assimilation and expulsion.
4. compare and contrast important theories on the nature of prejudice by Adorno and Allport.
5. demonstrate how slavery could be perceived as a cause of prejudice rather than a consequence of prejudice.
6. specify the nature of the association among status inequality, economic conflict and prejudice.
7. indicate how group identifiability and visibility impact prejudice and discrimination.
8. identify important educational and economic advances that have reduced status inequality among Japanese Americans.
9. state how geographic concentration, internal economic development and the development of a middle class function as mechanisms of racial and ethnic group mobility.
10. explain why Hispanic Americans are the largest minority group.

11. identify important within group patterns of inequality among Hispanic Americans.

12. explain why Cuban Americans report that they experience fewer instances of discrimination.

13. understand how the experiences of African American migrants from the rural South to the urban North could be compared to the experience of first generation immigrants to the U.S.

14. note important trends signifying reductions in status inequality between African Americans and whites and reductions in prejudice against African Americans.

15. identify factors that have had a lasting impact on African American mobility.

Chapter Outline

I. Intergroup Conflict
   A. Racial and Ethnic Inequality: The Basis of Intergroup Conflict
   B. Defining Race and Ethnicty
      1. Race addresses the social response to a group's perceived physical differences.
      2. Ethnicity addresses the social response to a group's perceived cultural differences.
   C. Outcomes of Intergroup Conflict
      1. Assimilation – conformity to dominant culture.
      2. Accommodation – emphasizing group similarities.
      3. Extermination – group annihilation.
      4. Expulsion – forced removal of a group.
      5. Segregation – separation of conflicting groups.

II. Theories of Prejudice
   A. Adorno's Authoritarian Personality Theory
      1. Persons accept only the norms and practices of their group.
      2. Anxiety is experienced when different norms and practices are encountered.
      3. Prejudice is employed as a defense mechanism to counter anxiety.
   B. Prejudice and Low Self-Esteem
   C. Education and Prejudice: An Inverse Association
   D. Allport's Theory of Contact
      1. Contact among competitive groups increases prejudice.
      2. Contact among cooperating groups reduces prejudice.
   E. Competitive Group Activities, Friendship Ties and Prejudice: The Sherif Studies

III. Status Inequality, Identifiability and Prejudice
   A. Slavery and the "American Dilemma"

1. Slavery generated racism and prejudice rather than racism and prejudice generating slavery.
2. Status inequality was maximized in the master-slave relationship.
3. U.S. history has been characterized by the paradox of democratic values and racist practices (the "American Dilemma").

B. Inequality and Prejudice: The U.S. Catholic - Protestant Experience
1. Prejudice toward Catholics diminished once economic parity with Protestants was attained.
2. Religious differences were markers or indicators of status conflict.

C. Economic Conflict and Prejudice
1. Minority groups can represent an economic threat to dominant group workers.
2. Bonacich identifies reasons minorities work for less.
   a. Low wages may represent an improvement in standard of living.
   b. Minority group members may not be aware they are being exploited.
   c. Racial and ethnic minorities may lack political power.
   d. Minority groups may be motivated by different economic concerns.
3. Dominant groups employ various mechanisms to reduce economic competition from minority groups.
   a. Immigration quotas can be established.
   b. Minorities may be restricted to low-wage employment opportunities.
   c. Minority groups may be scapegoated to mask class conflicts within the dominant group.

D. Identifiability and Prejudice
1. Competition between clearly identifiable groups stimulates harsher experiences of prejudice.
2. The "Okies" represented a low-wage minority group threat and were characterized by a distinctive rural life style and a Southern dialect.

IV. Status Equality, The Decline of Prejudice and Mechanisms of Rural and Ethnic Group Mobility

A. Status Equality Among Japanese Americans
1. Alien land laws, immigration restrictions and internment camps represent discriminatory acts directed toward Japanese Americans.
2. Japanese Americans enhanced their prospects for upward mobility by encouraging educational attainment and establishing small businesses and credit unions.
3. Highly educated Japanese Americans benefit from the post WWII economic expansion.
4. Economic parity with whites was attained by 1970.
5. High rates of Japanese American intermarriage reflect reductions in prejudice toward Japanese Americans.

B. Japanese Canadian Experience of Prejudice, Discrimination and Status Equality

C. Mechanisms of Racial and Ethnic Mobility
   1. Geographical concentration stimulates internal economic development.
   2. Successful internal economic development stimulates political development and the development of a strong middle class.

V. Hispanic Americans: The Largest U.S. Minority Group
   A. High Growth Rates: Immigration and Fertility Levels
   B. Geographic Concentration within Hispanic American Community
      1. Mexican Americans reside primarily in the Southwest.
      2. New York is home to a large Puerto Rican community.
      3. Cuban Americans are concentrated in Florida.
   C. Status Inequality and Hispanic Americans
      1. Hispanic Americans are twice as likely as the general population to be below the poverty line.
      2. Hispanic Americans are only half as likely as the general population to be a college graduate.
      3. As a sign of increasing cultural assimilation, Hispanic Americans are becoming more fluent in English and are developing more ties outside the Hispanic American community.
   D. Internal Stratification within the Hispanic American Community
      1. Cuban Americans have experienced the greatest degree of upward mobility with regard to educational attainment and income.
      2. Cuban Americans report experiencing the least degree of discrimination.

VI. African Americans: The Second Largest U.S. Minority Group
   A. Migration from Rural South to Urban North
   B. African American Migrants and First Generation Immigrants
   C. African American Educational and Income Gains: Mixed Findings
      1. The African American-white gap in educational attainment at the high school and college level has closed over the 1960-2000 period, but African Americans are still only two-thirds as likely as whites to have completed college.
      2. The African American-white earnings gap for working couple families has closed over the 1967-2000 period, but significant racial differences in family income exist when the earnings ratio is based on all households.
   D. Evidence of a Decline in Prejudice against African Americans
      1. Survey data suggest an increasing number of U.S. residents are living in integrated neighborhoods, interracial friendships are more common and interracial marriages are increasing.
      2. Firebaugh and Davis' analysis of 1972-1984 *General Social Survey* data indicates a general decline in prejudice toward African Americans with the most dramatic decline being experienced in the South.
      3. Schollaert and Smith's sociology of sports study indicates that the racial composition of a team does not impact fan attendance.
   E. Factors Exerting Lasting Impact on African American Experiences of

Prejudice, Discrimination and Mobility

1. The legacy of slavery institutionalized African American-white status inequality.
2. African Americans were denied basic political and economic freedom for an extended period of time.
3. African Americans are a highly identifiable and visible minority group and have been subject to harsh experiences of prejudice and discrimination.
4. African Americans are a large U.S. minority group that has at times been perceived as an economic threat.

Key Terms

Based on your reading of Chapter 11, you should be able to define and illustrate the key sociological concepts listed below. Page numbers are provided in parenthesis as reference points.

Interracial Conflict (287)
Intergroup Conflict (287)
Race (287)
Ethnic Groups (288)
Assimilation (288)
Accommodation (290)
Cultural Pluralism (290)
Extermination (290)
Expulsion (290)
Caste System (290)
Segregation (290)
Authoritarian Personality (292)
Allport's Theory of Contact (293)
Status Inequality (293)
The American Dilemma (296)
Markers (298)
Cultural Division of Labor (303)
Middleman Minorities (303)
Enclave Economy Theory (310)
Visibility (326)

Key Research Studies

Listed below are key research studies cited in Chapter 11. Familiarize yourself with the major findings of these studies. Page references are provided in parentheses.

Adorno: developed the theory of the authoritarian personality. Prejudice is a defense mechanism designed to control anxiety produced when people or groups characterized by a different set of norms and practices are encountered (292).

Allport: developed a theory of prejudice based on status inequality and contact. Prejudice increases when contact among groups competing for the same resources and opportunities increases (293-294).

The Sherif Studies: hostile stereotypes are generated, even among friends from different groups, when children engage in competitive rather than cooperative group activities (294).

Myrdal: wrote *An American Dilemma*, a classic study on U.S. race relations. This study highlights the paradox of democratic ideals and racial practices that has characterized U.S. history (296).

Bonacich: minority groups represent a potential economic threat to dominant group workers because minority groups can be a source of cheap labor (299-302).

Portes and colleagues: developed the enclave economy theory. This theory links geographical concentration, internal economic development and mobility (309-310).

de la Garza and colleagues: directed a national social survey of Hispanic Americans. Survey addresses many important issues within the Hispanic American community such as U.S. pride, English proficiency and experiences of discrimination (313, 316-317).

Lieberson: the migration experiences of African Americans from the rural South to the urban North should be compared to the experiences of first generation immigrants to the U.S. This comparison provides a context for a more accurate analysis of African American mobility (320).

Firebaugh and Davis: analyzed trends in prejudice based on the 1972-1984 *General Social Survey* data. Study documents a general decline in prejudice toward African Americans for the period with the most notable decline being observed in the South (323).

Schollaert and Smith: studied fan support of professional basketball. Study dispels racial myths and stereotypes by demonstrating that a team's racial composition does not impact fan support (324-325).

Info Trac Search Words

Enter these search terms to conduct more extensive investigations of key topics introduced in Chapter 11.

Ethnic Discrimination

View the articles listed and select one to investigate further. What social dimension of ethnic discrimination is addressed in the study selected? What solutions are proposed?

Segregation

Subdivisions: Analysis

Prejudices

Subdivisions: Social Aspects

Asian Americans

Subdivisions: Social Aspects

Select one of the studies cited. What social characteristics of the Asian American experience are addressed in your chosen study?

Discrimination in Sports

Multiple Choice

Answers and page references are provided at the chapter end.

1. Stark argues that intergroup conflict is based on
   a. racial and ethnic inequality.
   b. gender and age inequality.
   c. religious and political inequality.
   d. economic and kinship inequality.
   e. rural and urban inequality.

2. Sociologists define ethnicity in terms of the social response to
   a. physical differences.
   b. economic differences.
   c. cultural differences.
   d. political differences.
   e. all of the above

3. The primary ethnic ancestry of U.S. residents of European descent is
   a. English.
   b. German.
   c. Spanish.
   d. French.
   e. Italian.

4. During WWII many Jews met their death in German concentration camps. This is an example of
   a. assimilation.
   b. segregation.
   c. expulsion.
   d. extermination.
   e. accommodation.

5. According to the theory of the authoritarian personality, prejudice is
   a. influenced by high self-esteem.
   b. a defense mechanism.
   c. limited to intolerance of religious groups.
   d. a form of accommodation.
   e. restricted to ethnic group members.

6. Researchers have discovered that prejudice often decreases as
   a. self-esteem and education increase.
   b. self-esteem and education decrease.
   c. self-esteem increases and education decreases.
   d. self-esteem decreases and education increases.
   e. self-esteem and education remain stable over time.

7. Allport links prejudice with contact and
   a. high self-esteem.
   b. low self-esteem.
   c. status equality.
   d. achieved status.
   e. status inequality.

8. According to the Sherif studies, negative stereotypes are generated by
   a. cooperative tasks.
   b. competitive tasks.
   c. cooperative and competitive tasks.
   d. authoritarian leaders.
   e. authoritarian followers.

9. The major area of destination for African slaves being brought to the New World was
   a. British North America.
   b. the French Caribbean.
   c. Brazil.
   d. Spanish America.
   e. the Dutch Caribbean.

10. According to the 1790 Census, the state with the largest proportion of its population as slaves was
    a. North Carolina.
    b. Massachusetts.
    c. New York.
    d. South Carolina.
    e. Virginia.

11. The contradiction between democratic ideas and racist practices in the U.S. is known as
    a. the middleman minority.
    b. the enclave economy.
    c. identifiability.
    d. the "American dilemma."
    e. the occupational caste system.

12. When religious differences function as an indicator of an underlying status conflict between two groups, religious differences become
    a. markers.
    b. institutions.
    c. enclaves.
    d. cooperative tasks.
    e. cultural divisions of labor.

13. Bonacich argues that minority groups will accept low wages because
    a. low wages may be higher than the wages normally received.
    b. they may not be aware they are being exploited.
    c. they may lack political representation.
    d. they may only be looking for temporary employment.
    e. each factor is relevant.

14. In order to prevent minority group workers from competing with dominant group workers, immigration quotas may be set. This is an example of
    a. an occupational caste system.
    b. exclusion.
    c. the cultural division of labor
    d. middleman minorities.
    e. markers.

15. Farming was the main occupation of many first generation Japanese immigrants. This is an example of the following concept.
    a. geographic concentration
    b. extermination
    c. cultural division of labor
    d. middleman minorities
    e. expulsion

16. The Utes are a dark skinned Native American group that was hated by other Native American groups and whites. This example shows the interrelationship between prejudice and
    a. status equality.
    b. slavery.
    c. identifiability.
    d. self-esteem.
    e. accommodation.

17. The "yellow peril" referred to the
    a. immigration of Puerto Rican Americans to New York.
    b. immigration of Cuban Americans to Florida.
    c. migration of Native Americans from reservation to reservation.
    d. migration of African Americans from the rural South to the urban North.
    e. immigration of Japanese Americans to Hawaii and the West Coast.

18. Which of the following statements about the Japanese American experience in the U.S. is false.
    a. Japanese Americans were less likely than whites to be enrolled in school prior to WWII.
    b. During WWII a significant number of Japanese Americans were placed in internment camps.
    c. Japanese Americans created their own credit unions early.
    d. Migration from Japan was prohibited in 1924.
    e. Californians passed laws prohibiting Japanese land ownership.

19. Since 1970, Japanese American earnings have
    a been lower than white earnings.
    b. been equal to white earnings.
    c. been equal to or lower than white earnings.
    d. been equal to or higher than white earnings.
    e. not been formally measured.

20. Intermarriage rates among Japanese Canadians are
    a. below 1%.
    b. approximately 5%.
    c. approximately 10%.
    d. at 25%.
    e. above 50%.

21. Racial, ethnic and religious minorities often live in segregated urban enclaves. This is known as
    a. the "American dilemma."
    b. geographical concentration.
    c. the occupational caste system.
    d. the status inequality hypothesis.
    e. the middleman minority.

22. During the first half of the twentieth century, Japanese Americans were successful in establishing their own farms, small businesses and credit unions. This is an example of
    a. the creation of an enclave economy.
    b. intergroup conflict.
    c. the middleman minority.
    d. the legacy of Japanese slavery.
    e. the lack of a Japanese homeland.

23. Presently the largest racial or ethnic minority group in the U.S. is
    a. Hispanic Americans.
    b. African Americans.
    c. Japanese Americans.
    d. Chinese Americans.
    e. Native Americans.

24. The Mexican American population primarily resides in
    a. Florida.
    b. New York.
    c. the Midwest.
    d. the Southwest.
    e. the Southeast.

25. Which one of the following statements concerning differences within the Hispanic American community is true.
    a. Cuban Americans are most likely to live in poverty.
    b. Mexican Americans are most likely to be employed in high status occupations.
    c. Puerto Rican Americans have attained the highest levels of educational achievement
    d. Cuban Americans are less likely to feel that they have been discriminated against.
    e. Each statement is true.

26. An increase in Hispanic American intermarriage, a decline in Spanish fluency, and the purchasing of homes outside Hispanic neighborhoods reflect a greater degree of
    a. prejudice.
    b. discrimination.
    c. assimilation.
    d. segregation.
    e. expulsion.

27. Prior to 1940, the majority of African Americans resided in the
    a. urban North.
    b. rural South.
    c. urban West.
    d. urban South.
    e. rural North.

28. Which of the following statements is true.
    a. The African American-white gap in educational attainment is increasing.
    b. The income gap between African American and white working couple households is increasing.
    c. African Americans are nine times more likely to live in segregated neighborhoods.
    d. The number of interracial friendships and marriages is declining.
    e. Each of these statements is false.

29. Firebaugh and Davis discovered that the most dramatic decline in prejudice against African Americans has occurred in the
    a. North.
    b. East.
    c. South.
    d. West.
    e. The decline in prejudice has been the same in each region.

30. Stark argues that African American experiences of prejudice, discrimination and mobility have been influenced by
    a. the legacy of slavery.
    b. the lack of a homeland.
    c. African Americans' high degree of visibility.
    d. dominant group perceptions of African Americans as economic competitors.
    e. all of the above

True/False

Answers and page references are provided at the chapter end.

1. Sociologists generally associate race with a society's response to a group's noticeable cultural characteristics.

2. Extermination, expulsion and segregation are three possible outcomes of intergroup conflict.

3. Researchers have observed that prejudice increases as self-esteem and education increases.

4. According to Allport's theory of contact, prejudice decreases as contact among groups cooperating to pursue common goals increases.

5. Stark argues that racist attitudes led to the institution of slavery in the U.S. rather than the institution of slavery promoting racial beliefs.

6. Women and minorities are often concentrated in low-paying data entry jobs. This is an example of the cultural division of labor concept.

7. Prejudice tends to become more hostile as a group's identifiability and visibility decreases.

8. Immigration restrictions and internment camps are examples of prejudice and discrimination directed toward Japanese Americans.

9. Compared to other Hispanic American groups, Puerto Rican Americans have experienced the greatest degree of upward mobility.

10. Schollaert and Smith's study of professional basketball indicates that a team's racial composition does not have a major impact on fan attendance. White fans will support teams comprised of a large number of African American players.

Short Answer Questions

These short answer questions are provided to test your knowledge and understanding of the basic sociological concepts presented in Chapter 11. Page references for answers are included at the chapter end.

1. How do sociologists generally distinguish race and ethnicity?

2. Identify five possible outcomes of intergroup conflict.

3. Compare and contrast the authoritarian personality and contact theories of prejudice.

4. What is the "American Dilemma?"

5. Describe the nature of the association between status inequality and prejudice.

6. According to Bonacich, why are minority groups willing to work for less?

7. Provide an example of the cultural division of labor.

8. How does geographic concentration, internal economic development and the development of a middle class enhance racial and ethnic mobility?

9. What is the nature of status inequality within the Hispanic American community?

10. State the major findings of the Firebaugh and Davis study on prejudice against African Americans.

Essay Questions

These questions are designed to test your understanding of key sociological concepts presented in Chapter 11 and your ability to apply these insights to concrete situations.

1. Critically evaluate the following statement. "Urban Renewal Programs are examples of racial and ethnic group expulsion."

2. Demonstrate how prejudice can function as both an independent variable and a dependent variable. Provide concrete examples from everyday life.

3. How could Spanish be perceived as an example of a cultural marker in contemporary U.S. society.

4. Hispanic Americans have experienced between group and within group status inequality. Provide evidence to support this claim.

5. Firebaugh and Davis note that the decline in prejudice against African Americans has been more rapid in the South. Identify testable factors that could be associated with this dramatic change.

Answers

Multiple Choice

1. a (287)
2. c (287-288)
3. b (288)
4. d (290)
5. b (292)
6. a (292)
7. e (293)
8. b (294)
9. c (295)
10. d (296)
11. d (296)
12. a (298)
13. e (299-302)
14. b (302)
15. c (303)
16. c (285-286, 303-304)
17. e (304)
18. a (304-306)
19. d (306)
20. e (306)
21. b (309)
22. a (309-310)
23. a (312)
24. d (312)
25. d (314-317)
26. c (317)
27. b (318)
28. e (320-323)
29. c (323)
30. e (323-327)

True/False

1. F (287-288)
2. T (290)
3. F (292)
4. T (293-294)
5. F (294-297)
6. T (303)
7. F (303-304, 326)
8. T (304-306)
9. F (313-315)
10. T (324-325)

Short Answer

1. (287-288)
2. (288-290)
3. (292-294)
4. (296)
5. (297-303)
6. (299-302)
7. (302-303)
8. (308-312)
9. (313-317)
10. (323)

CHAPTER 12

GENDER AND INEQUALITY

Extended Abstract

Women and men do not share equal power. This is demonstrated clearly in non-industrial societies where almost nine times out of ten, political leaders are always men. Gender inequality is less pronounced in industrial societies, but is still evident. Gender Power Ratios indicate that Norwegian women have 84 % as much power as men, but for U.S. and Japanese women, this figure falls to 74 % and 52 % respectively. Why do these gender differences in power exist? Stark claims the key lies in variations in the sex ratio, the supply of males to females.

According to a theory developed by Guttentag and Secord, gender roles and gender relationships are influenced by the sex ratio. Gender roles and relationships are shaped by social structures. Where there is a surplus of men, women lack power, and where women outnumber men, the status of women is higher. Sex ratio imbalances can be created by a variety of factors such as geographic mobility which favors males, female infanticide, health and diet, differential life expectancy, warfare and various sexual practices.

Before specifying the association between the sex ratio and power relationships, Stark describes gender relationships in two classical Greek city-states, Athens and Sparta. Athens is characterized by a high sex ratio (oversupply of males) while Sparta is characterized by a low sex ratio (oversupply of women). The practice of female infanticide appears to have contributed to the Athenian sex ratio imbalance. Given the oversupply of men, women's status was low. Women received little formal education, married early, did not enjoy legal privileges, were regarded as male property and were subject to protective sexual norms. Spartan women were in greater supply and were more powerful. Spartan women often managed estates since men were regularly involved in warfare. Women received formal education, were able to own land, possessed certain rights in cases of divorce, married at a later age and were permitted more sexual freedom.

Turning to a discussion of the sex ratio - power association, Stark introduces the following power concepts, dyadic power, power dependence and structural power. When women are in short supply, they are able to maximize their power within a dyadic association. In response to this experience of power dependency, men, who are in greater supply, utilize their strength in numbers to establish norms and social structures that will benefit men. Women's power outside the dyad is limited, and men and women will be socialized to conform to gender role expectations favoring males.

The claims of the Guttentag and Secord sex ratios and sex roles theory are tested in the remaining portions of the chapter. The theory is utilized to gain a better understanding of the rise of the feminist movement in the U.S., women's labor force participation, the sexual revolution and changing African American family structures.

For whites, females have outnumbered males since 1950, but for African Americans, females have outnumbered males since 1840. Sex ratios vary also among age groups. When the sex ratios for the ages that men and women typically marry are compared, women tend to outnumber men. This phenomenon is known as a "marriage squeeze" and is more pronounced within the African American community. According to Guttenberg and Secord, when women lack dyadic power and experience power dependency, they seek ways to structure society so as to maximize the power reflected in their strength in numbers. Two responses designed to enhance women's status and reduce gender inequality are the "Woman Movement" and the "Feminist Movement." The "Woman Movement" originated and gained great support in the Northeast where historically women had been in greater supply. The "Woman Movement" ended once women gained the right to vote, but the "Feminist Movement" soon replaced this movement.

The increased labor force participation of women is another consequence of the shortage of men. During the nineteenth century, women's labor force participation in the eastern states was high since many young men were migrating west. Presently, six out of ten women are employed outside the home, but gender differences in earnings persist. Taeuber and Valdisers note that women are often concentrated in lower paying jobs, experience more career interruptions and tend to acquire training in fields that pay less.

The sexual revolution can also be understood as a consequence of the oversupply of women. Within this context women have the power to demand sexual freedom, and since men have potential sexual access to more women, restrictive sexual norms are relaxed. As women gain greater sexual freedom, the potential for sexual exploitation increases also.

Finally, changes in the African American family structure may be linked to sizeable sex ratio imbalances. Within the African American community, a severe shortage of men exists. This imbalance has been attributed to high rates of male fetal death and infant mortality, and to high rates of mortality among young African American men. Since interracial marriage is lower among African American females, and since the supply of African American men of marriageable age is low, marriage opportunities are low. Given the short supply of men, African American women may be less likely to marry, more likely to conceive children outside of marriage and more involved in the labor force. Recent research indicates that the correlation between variations in the African American sex ratio and African American single parent households is strong. Similar findings have been observed among Puerto Rican Americans where the sex ratio is also low.

Key Learning Objectives

After a thorough reading of Chapter 12, you should be able to:

1. verify the low status of women in non-industrial societies.

2. state the association among industrialization, standard of living and status of women.

3. indicate how the sex ratio is calculated and interpreted.

4. note important variations in the U.S. sex ratio at the national and regional level over time.

5. state the major tenets of the Guttentag and Secord sex ratios and sex roles theory.

6. identify factors associated with the creation of sex ratio imbalances.

7. compare and contrast the status of women in two classical Greek city-states that were characterized by varying sex ratios.

8. distinguish dyadic power, power dependency and structural power.

9. note how an unbalanced sex ratio may create a "marriage squeeze."

10. demonstrate how the sex ratios and sex roles theory may provide insight into the origin of gender-based social movements.

11. identify some of the major concerns of the "Woman Movement" and the "Feminist Movement."

12. indicate how sex ratio imbalances may impact women's labor force participation.

13. state the association between variations in the sex ratio and changing sexual norms.

14. indicate how shortages in the supply of African American men could impact the African American family.

15. identify important variations in the sex ratio within the Hispanic American community.

Chapter Outline

I. Dimensions of Gender Inequality
    A. Gender Inequality in Non-Industrial Societies
        1. Women generally lack political authority.
        2. Husbands tend to dominate their wives.
    B. Gender Inequality in Industrial Societies
        1. The status of women generally increases as industrialization and standard of living increases.
        2. The Gender Power Ratio is an index measuring women's status based on women's political involvement, labor force involvement and earnings.
        3. Significant variations in women's power exist in industrial nations.

II. The Guttentag and Secord Sex Ratios and Sex Roles Theory
    A. The Sex Ratio: The Supply of Men to Women
    B. National and Regional Variations in the Sex Ratio: 1790, 1880 and 1990.
    C. The Status of Women and the Sex Ratio
        1. Women are more powerful when there is an oversupply of women.
        2. Men are more powerful when there is an oversupply of men.
    D. Factors Contributing to Sex Ratio Imbalances
        1. Geographic mobility creates an oversupply of males in areas of destination and leaves an oversupply of females in areas of origin.
        2. Female infanticide limits the supply of women.

3. Health, diet and differential life expectancy impact the supply of men and women.
4. Warfare and sexual practices influence the supply of males.

E. The Sex Ratio and the Status of Athenian and Spartan Women
1. Ancient Athens is an example of a high sex ratio society where female infanticide was practiced, women received little education, women were treated as property and sexual norms were strict.
2. Ancient Sparta is an example of a low sex ratio society where male and female infanticide was practiced, women received formal educational training, women were able to own land and sexual norms were more lax.

F. The Sex Ratio and Power Relationships
1. Within a dyadic relationship, the gender in the shorter supply has more power (dyadic power) and the gender in greater supply experiences power dependency.
2. Outside the dyadic relationship, the gender in greater supply has more power (structural power) and will create societal arrangements that enhance that gender's status.

III. Testing the Sex Ratios and Sex Roles Theory

A. The Oversupply of Women in the U.S.
1. There has been an oversupply of white women in the U.S. since 1950.
2. There has been an oversupply of African American in the U.S. since 1840.
3. The "marriage squeeze" signals an oversupply of women of marriageable age.

B. The Sex Ratios and Sex Roles Theory and Gender Based Social Movements
1. The "Woman Movement" originated in the Northeast where women had historically been in greater supply.
2. The "Woman Movement" declined once women were granted suffrage.
3. The "Feminist Movement" followed the "Woman Movement" and addressed more issues concerning gender inequality.

C. A Low Sex Ratio and Greater Women's Labor Force Participation
1. Approximately six in ten women are involved in full-time work outside the home.
2. Women are concentrated in lower-paying occupations, experience more career interruptions and often do not choose majors that lead to high-income jobs.

D. The Sexual Revolution and the Surplus of Women
1. Women tend to enjoy more sexual freedom when there is a surplus of women.
2. The sexual exploitation of women is a potential outcome of women's sexual freedom.

E. The Shortage of Men and the African American Family
1. African American men are in short supply due to high rates of fetal and infant mortality and to high rates of young adult mortality.

2. The shortage of males of marriageable age has created a "marriage squeeze" within the African American community.

4. Interracial marriage among African American females is low.

5.

4. Given the sex ratio imbalance, African American women have fewer marriage choices, are more likely to conceive a child outside of marriage and are more involved in the labor force.

5. Recent research indicates that the correlation between variations in the African American sex ratio and single parent households is strong.

F. Variations in the Sex Ratio and Hispanic Americans

1. As a result of higher rates of male immigration Mexican American men tend to outnumber Mexican American women.

2. A shortage of males of marriageable age has created a "marriage squeeze" within the Puerto Rican American community.

Key Terms

Based on your reading of Chapter 12, you should be able to define and illustrate the key sociological concepts listed below. Page numbers are provided in parentheses as reference points.

Gender Power Ratio (332)
Sex Ratio (334)
Infanticide (339)
Dyadic Power (344)
Power Dependency (344)
Structural Power (345)
Marriage Squeeze (347)
Woman Movement (348)
Feminism (350)
Spinster (351)
Illegitimacy Ratio (359)

Key Research Studies

Listed below are key research studies cited in Chapter 12. Familiarize yourself with the major findings of these studies. Page references are provided in parentheses.

Standard Cross Cultural Sample: source for data on gender inequality in non-industrial Societies (331, 333).

Gender Power Ratio: index developed by United Nations staff to measure gender inequality. Measure is based on women's political and economic involvement (332).

Guttentag and Secord: developed theory linking sex ratios and sex roles. The status of women and the sex role vary inversely (337-339).

Taeuber and Valdisera: studied gender differences in income. Findings indicate that women tend to be concentrated in lower-paying occupations, are more likely to experience career interruptions and are more likely to acquire educational training in fields that do not lead to higher-paying jobs (354-355).

Fossett and Kiecolt: observed a strong correlation between variations in African American sex ratios and single parent African American households. More women marry as the supply of men increases (359).

Info Trac Search Words

Enter these search terms to conduct more extensive investigations of key topics introduced in Chapter 12.

Gender Inequality
Review the studies cited and select one for more extensive study. Identify the aspects of gender inequality addressed by the study. How extensive is the inequality and what actions are proposed to reduce the inequality?

Gender and Power
Select one of the studies listed. How is women's power being defined? What factors are identified as influencing women's power?

Sex Ratio
Subdivisions: Research

Feminism
Subdivisions: Social Aspects

African Americans
Subdivisions: Mortality

Multiple Choice

Answers and page references are provided at the chapter end.

1. Each of the following statements about gender relations in non-industrial societies is true except
   a. political leaders are always male in the majority of non-industrial societies.
   b. wife beating is rare in non-industrial societies.
   c. the general belief is that men should dominate their wives.
   d. rape is frequent in non-industrial societies.
   e. women have less control over their marital and sex lives.

2. The status of women increases in societies as
   a. industrialization increases.
   b. industrialization decreases.
   c. the standard of living increases.
   d. a and c
   e. b and c

3. The U.S. gender power ratio is .74. This means that
   a. men have 74 percent as much power as women.
   b. women have 74 percent as much power as men.
   c. men have 26 percent as much power as women.
   d. women have 26 percent as much power as men.
   e. men and women have close to an equal amount of power.

4. The sex ratio is usually expressed as the
   a. number of males per 100 females.
   b. number of females per 100 males.
   c. number of males per 1,000 females.
   d. number of females per 1,000 males.
   e. none of the above

5. The 1790 Census revealed that there was a surplus of women in these three states.
   a. Georgia, South Carolina and North Carolina
   b. North Carolina, Virginia and Maryland
   c. Maryland, Pennsylvania and New Jersey
   d. New Jersey, New York and Connecticut
   e. Massachusetts, Rhode Island and Connecticut

6. Data from the *Nations of the Globe* data set indicate that women are more likely to participate in the labor force as the sex ratio decreases. This is an example of a/an
   a. positive correlation.
   b. spurious correlation.
   c. negative correlation.
   d. structural correlation.
   e. exchange correlation.

7. The killing of male or female infants is known as
   a. the sex ratio.
   b. the gender power ratio.
   c. infanticide.
   d. dyadic power.
   e. illegitimacy.

8. A major factor creating a shortage of African American males has been
   a. a significant migration of African American males outside the U.S.
   b. an increase in female infanticide.
   c. the increase in African American male life expectancy.
   d. limited military involvement of African American males.
   e. a high rate of male fetal deaths.

9. Women in ancient Athens
   a. greatly outnumbered men.
   b. received a significant amount of education.
   c. married early.
   d. were able to own property.
   e. often wore dresses that were short and sleeveless.

10. Spartan women
   a. could own property.
   b. received as much education as men.
   c. married late.
   d. received gymnastic training.
   e. all of the above

11. Guttentag and Secord's sex ratios and sex roles theory incorporates aspects of
   a. exchange theory and social evolutionary theory.
   b. conflict theory and exchange theory.
   c. control theory and conflict theory.
   d. functionalist theory and social evolutionary theory.
   e. social evolutionary theory and rational choice theory.

12. In a society characterized by an oversupply of women, men will have more power in a two-member group. This is known as
   a. the gender power ratio.
   b. structural power.
   c. power independence.
   d. dyadic power.
   e. feminism.

13. The norm, "brides shall be virgins," illustrates
   a. structural power.
   b. feminism.
   c. a low sex ratio.
   d. a high gender power ratio.
   e. illegitimate power.

14. When women are in greater supply, they
   a. have more dyadic power.
   b. experience a greater degree of power dependency.
   c. possess more structural power.
   d. a and b
   e. b and c

15. Presently the sex ratio for African Americans
   a. is the same as the white sex ratio.
   b. cannot be calculated.
   c. is lower than the white sex ratio.
   d. is higher than the white sex ratio.
   e. none of the above

16. Which of the following statements is correct?
   a. The African American sex ratio fell below 100 before the white sex ratio did.
   b. The African American and white sex ratios fell below 100 at the same time.
   c. The white sex ratio fell below 100 before the African American sex ratio did.
   d. Sex ratio data for African Americans has been available as long as it has for whites.
   e. During most of the nineteenth century, the African American sex ratio was higher than the white sex ratio.

17. Until recent times, the surplus of males in Mexico was attributed primarily to
   a. the poor diet of females.
   b. female infanticide.
   c. immigration.
   d. high rates of female mortality associated with childbirth.
   e. male infanticide.

18. In Canada the surplus of men has been attributed primarily to
   a. lower female life expectancy.
   b. immigration.
   c. female infanticide.
   d. poor maternal health care.
   e. poor dietary practices of women.

19. In most societies women
   a. choose to live with another man rather than get married.
   b. form single parent households rather than two parent households.
   c. marry someone their own age.
   d. marry a man who is younger than they are.
   e. marry a man who is older than they are.

20. The "marriage squeeze" is associated with
    a. an oversupply of women of marriageable age.
    b. the pressure for women to marry at a young age.
    c. the pressure for men to marry at a young age.
    d. the age of marriage, divorce and remarriage.
    e. the famed practice of "mail order" brides.

21. The earliest women's rights movement was known as the
    a. Gender Power Movement.
    b. Feminine Mystique.
    c. Feminist Movement.
    d. ERA.
    e. Woman Movement.

22. The "Woman Movement" began in the __________ where historically there had been a shortage of __________.
    a. West; men
    b. West; women
    c. Northeast; men
    d. Northeast; women
    e. South; men

23. The Nineteenth Amendment guaranteed women the right
    a. to own property.
    b. vote.
    c. hold public office.
    d. receive equal pay for equal work.
    e. marry outside their racial or ethnic group.

24. The two forerunners of the Feminist Movement were
    a. the Civil Rights Movement and the Suffrage Movement.
    b. the Woman Movement and the Labor Movement.
    c. the Civil Rights Movement and the Labor Movement.
    d. the Suffrage Movement and the Woman Movement.
    e. the Gay Rights movement and the Woman Movement.

25. The early Feminist Movement
    a. began as a small, elite movement.
    b. attracted women involved with more radical political groups.
    c. attracted women involved in the labor movement.
    d. involved young female intellectuals.
    e. all the above

26. By 2000 __________ of all U.S. women were employed full-time outside the home.
   a. 25%
   b. 33%
   c. 40%
   d. 50%
   e. 60%

27. Taeuber and Valdisera argue that gender differences in earnings may be attributed to the fact that
   a. women are concentrated in lower-paying occupations.
   b. the average working woman is much older than the average working man.
   c. men change jobs more often than do women.
   d. women generally have more job seniority.
   e. a man is more likely to experience career interruption.

28. When there is a surplus of women
   a. women enjoy more sexual freedom.
   b. women enjoy less sexual freedom.
   c. women are more likely to be sexually exploited and treated as sex objects.
   d. a and c
   e. b and c

29. The large number of female-headed households among African Americans is
   a. part of the legacy of slavery.
   b. linked to the short supply of African American men.
   c. has been the dominant household structure for over a hundred years.
   d. linked to the short supply of African American women.
   e. a result of high levels of African American female mortality.

30. Puerto Rican American
   a. women are more likely to face a shortage of men.
   b. men are more likely to face a shortage of women.
   c. families are more likely to be headed by a single parent.
   d. a and c
   e. b and c

True/False

Answers and page references are provided at the chapter end.

1. In the majority of non-industrial societies, women and men have equal political rights.

2. A sex ratio over 100 means that there are more females in the society than males.

3. The westward migration of young men during the nineteenth century resulted in a surplus of women in the eastern states and a surplus of men in the western states.

4. According to the Guttentag and Secord theory of sex ratios and sex roles, the status of women increases where men outnumber women.

5. Stark maintains that female infanticide is the major cause of sex ratio imbalances.

6. Classical Athens was characterized by an oversupply of men. Women were typically denied the right to own land and married early.

7. When men are in short supply, they also have less dyadic power.

8. The "Woman Movement" originated in the Northeast where men had historically been in short supply.

9. According to the Guttentag and Secord sex ratios and sex roles theory, the potential for the exploitation of women increases as the status of women increases.

10. Interracial marriage rates are higher among African American women than African American men.

Short Answer Questions

These short answer questions are provided to test your knowledge and understanding of the basic sociological concepts presented in Chapter 12. Page references for answers are included at the chapter end.

1. How does the status of women vary in industrial and non-industrial societies?

2. What is the sex ratio, and how is it interpreted?

3. Identify six factors that can lead to sex ratio imbalances.

4. Ancient Athens had too many men and Ancient Sparta had too many women. How did this effect women's status in each location?

5. Specify the interrelationship among dyadic power, power dependency and structural power.

6. What are the major variations in African American and white sex ratios over the past two centuries?

7. What is the "marriage squeeze?"

8. Did the "Woman Movement" and the early "Feminist Movement" attract the same audience?

9. Gender differences in income persist even though more women are participating in

the labor force. What factors are associated with gender differences in income?

10. Why is there a shortage of men within the African American community, and what impact is this having on the African American family?

Essay Questions

These questions are designed to test your understanding of key sociological concepts presented in Chapter 12 and your ability to apply these insights to concrete situations.

1. How do increasing industrialization and modernization improve women's status?

2. Gender roles are becoming more androgynous. This means that traditional gender-specific behavioral characteristics are becoming less gender-specific. How would the Guttentag and Secord theory account for the rise in androgyny?

3. Why did the surplus of African American women occur more than one hundred years before the surplus of white women?

4. According to the Guttentag and Secord theory, why would pornography and other forms of sexual exploitation increase as the status of women increases?

5. In trying to identify some of the major social factors associated with gender inequality, why would it be important to look at the interaction between gender, race and ethnicity?

Answers

Multiple Choice

1. b (331, 333)
2. d (332)
3. b (332, 334)
4. a (334, 336)
5. e (336)
6. c (339)
7. c (339)
8. e (341)
9. c (342-344)
10. e (342-344)
11. b (344-345)
12. d (344)
13. a (345)
14. e (344-345)
15. c (346)
16. a (346)
17. d (346)
18. b (346)
19. e (346)
20. a (347)
21. e (348-349)
22. c (348, 350)
23. b (350)
24. d (350)
25. e (351)
26. e (351)
27. a (354-355)
28. d (355, 357)
29. b (358)
30. d (369-360)

True/False

1. F (331, 333)
2. F (334, 336)
3. T (338)
4. F (334, 336-339)
5. T (339)
6. T (342)
7. F (344)
8. T (350)
9. T (355, 357)
10. F (358)

Short Answer

1. (331-334)
2. (334, 336)
3. (339-342)
4. (342-344)
5. (344-346)
6. (346-347)
7. (347)
8. (348-350)
9. (354-355)
10. (358-359)

CHAPTER 13

THE FAMILY

Extended Abstract

Two themes have dominated the sociological study of the family. One concerns the universal nature of the family while the other addresses the decline in the family's importance in the modern world. Although the family may be a universal social phenomenon, tremendous variations in family structure exist. Can a general, comprehensive definition of the family be provided? Is the family's importance declining? What is the reference point for measuring the decline? Was life in non-industrial societies necessarily better? These issues are addressed in this chapter.

Murdock offers a classic definition of the family. His definition stressed common residence, economic cooperation, reproduction and child rearing. The family would also include adults of both sexes and children. However, this definition is too restrictive. Would communal living be included? What about childless households or single parent households? A more concise definition is offered by Reiss who maintains that the family is a kinship-based group whose primary function is child rearing. Here again, this definition is too restrictive since some couples are unable to have children and others choose not to have children. So, what may be regarded as some of the primary characteristics of non-industrial families, and what are some of the primary functions of the family in more contemporary societies?

Data from the *Standard Cross-Cultural Sample* indicate that in the majority of non-industrial societies, men had multiple wives and could easily divorce their wives while couples generally lived near the groom's family. In many instances, men did not help with domestic tasks and the nuclear family structure was atypical. Since extensive variation in family life and family structure exists for non-industrial and industrial families, sociologists have found comparative studies of commonly recognized family functions to be helpful. These functions include sexual gratification, economic support and emotional support. Data from the *World Values Survey* are presented to demonstrate how societies vary in their perception of these functions. Norms regulating sexual behavior are found in all societies. In some instances, the norms are more restrictive than in others, and spouses may not share the same sexual attitudes. In many societies, the family is the primary source of economic support for its members. This is particularly the case with dependents such as children. In more industrialized societies, economic support of family members will be shared with other institutions through work-related pension programs and public assistance. The family is a primary group, and often provides needed emotional support for its members. However, satisfaction with family life growing up and perceived closeness of spouses varies among industrial and industrializing societies.

Shorter's study of extended family life in seventeenth century Europe challenges the assumptions about the strong, close ties associated with traditional extended families. Shorter argues that these traditional European extended families often included non-family members and female-headed households were not uncommon. Infant mortality rates were high, life expectancy was low, and many children left home early to participate in apprentice programs. Consequently, these extended families were quite small. Many homes contained only one room. Since that space could be shared with non-family members and livestock, crowding was extensive and privacy was minimal. Available medical records indicate that children were victims of neglect, particularly during harvest season. Funeral and burial practices reflected weak parent-child ties as funerals for young children were at times not attended by parents and deceased children were simply discarded. Illegitimate children were often abandoned. Parent-child ties with older children were weak since many left home early to serve as apprentices. Spouse ties were weak as marriages were based on economic arrangements rather than emotional attachments. On the other hand, adult peer group ties were strong. These groups provided the needed emotional support.

Stark argues that changes in the strength of family ties are an outcome of modernization and industrialization. As more economic opportunities were created for workers, family members did not have to place as much emphasis on inheritance. As workers became more affluent, they could afford larger houses. Spouse ties could be based more on strong emotional attachments rather than economic arrangements. More emphasis was placed also on parenting skills. Modernization has been associated with reducing the quantity of close-kin ties, but not necessarily the quality of these relationships. Today, families are producing fewer children. This limits the number of close-kin ties that can be developed; however, social survey data indicate that contact with mothers and relatives is maintained on a fairly frequent basis.

The remainder of the chapter is devoted to a discussion of important issues impacting U.S. family life. These issues include divorce, cohabitation, one-parent families, remarriage and household responsibilities. People divorce for many reasons, such as loss of emotional satisfaction and sexual attraction, extramarital affairs, high expectations and the liberalization of divorce laws. However, high divorce rates may also suggest that more satisfying marriages are desired and considered important. Divorce rates are high, but remarriage rates are high. Also, most married persons do not divorce. Many persons who have divorced continue to remarry and divorce. It is difficult to generate accurate measures of divorce. Some measures are based on the total population, others involve calculating divorce-marriage ratios and some measures have been based on final outcomes. Measurement and interpretation problems are associated with each divorce indicator.

A comprehensive cross-national study on divorce by Trent and South is provided as a representative case study on divorce. Four hypotheses are tested with data from sixty-six countries. The first hypothesis associates an increase in divorce with a decline in the importance of the family that accompanies increasing urbanization and economic development. The second hypothesis links an increase in divorce with increased female labor force participation. The third hypothesis tests the Guttentag and Secord sex ratios and sex roles theory. Men are less likely to divorce when the supply of women is low. The final hypothesis addresses cultural influences. Given the Catholic Church's ban on divorce, divorce varies inversely with the proportion of the population Catholic. Support for the first three hypotheses is provided by the data.

The number of U.S. couples living together prior to marriage is increasing. Some argue that living together enables people to reduce the number of future unions that would be unsuccessful. Others maintain that living together prior to marriage reflects an unwillingness to make lasting commitments. A Swedish study on cohabitation indicates that the divorce rate among cohabitating couples is high. Also, the number of one-parent families is increasing. This may be the result of divorce, death of a spouse or the pregnancy of a single mother. Births to unmarried mothers have increased significantly in the U.S. since 1960. One-parent families face important challenges. Primary among them are the need for more income and more time to meet the demands of everyday family life.

The last two topics addressed are remarriage and household responsibilities. In a study of conjugal careers or remarriages, Jacobs and Furstenberg observed that the social status of a woman's second husband is similar to the current economic status of her first husband. Differences in status between the first and second husband occur when the woman is significantly older when she remarries or when she has young children from a previous marriage. In the first instance, the second husband's status may be either higher or lower, but in the second instance, the second husband's status is generally lower than the first husband's. Likewise, White and Booth discovered that the odds of divorce in a remarriage increase significantly if both parties have been previously divorced and if either or both partners bring children from a previous marriage into the new home. They also discovered that stepchildren leave home at an earlier age. In related research, White and Edwards note that parental marital happiness and life satisfaction increase once the last child leaves home, but that happiness is contingent upon parents being able to maintain close contact with their children. Finally, cross-national comparisons of data on traditional household responsibilities indicate that many household tasks have remained gender-specific even though women's labor force participation has increased.

Key Learning Objectives

After a thorough reading of Chapter 13, you should be able to:

1. critically evaluate general definitions of the family.
2. identify major patterns of family life in non-industrial and industrial societies.
3. distinguish key needs addressed by the family as a social institution.
4. explain why seventeenth century European extended families were small in size.
5. challenge traditional stereotypes characterizing family ties within extended family networks.
6. assess the impact of modernization on spouse ties and other kinship ties.
7. discuss how higher divorce rates could indicate that more satisfying marriages are desired and considered important.
8. identify problems associated with measuring the divorce rate.
9. note current trends in divorce among U.S. women by race, ethnicity, age at first marriage, family income, religion and parents' conjugal career.
10. assess the impact of modernization, female labor force participation, the sex ratio and cultural belief systems on divorce.
11. state varying interpretations of the impact of living together on marriage and divorce.

12. identify specific needs encountered in one-parent families.

13. evaluate the impact of remarriage on mate selection, parent-child relationships and divorce.

14. identify some of the major issues associated with the "empty nest syndrome," and the "postlaunch honeymoon."

15. comment on the extent that household tasks have remained gender-specific even though women's labor force participation has increased.

Chapter Outline

I. The Family as a Social Institution
   A. A Problem of Definition: Murdock and Reiss
      1. The family is a social group comprised of adults of both sexes and children addressing such needs as economic cooperation, reproduction and child care (Murdock).
      2. The family is a kinship-based group whose primary responsibility is childrearing (Reiss).
   B Patterns of Family Life
      1. Non-industrial families are characterized by polygamy, male power and extended families.
      2. Contemporary cross-national data reflect variations in the perception of common family functions like norms governing sexual behavior, the importance of economic well-being and care of dependents and the provision of emotional support.
      3. Shorter's study of seventeenth century European extended families challenges extended family stereotypes.
         a. European extended families were small due to high infant mortality, low life expectancy and children leaving home at an early age to serve as apprentices.
         b. Crowding was common, privacy was limited and children often suffered from neglect.
         c. Parent-child and spouse ties were weak; whereas, adult peer group ties were strong.
      4. Modernization brought about improvements in the quality of family life.
         a. As wage labor reduced the dependence on inheritance, marriage was transformed from an economic arrangement to an association based on stronger emotional attachments.
         b. Although modernization is associated with smaller family networks, continued contact with family members and relatives is maintained in contemporary industrial societies like the U.S.

II. Issues Facing Contemporary U.S. Families

A. Divorce: Satisfaction or Dissatisfaction with Marriage?
   1. High divorce rates may signal high marital expectations since remarriage rates are high.
   2. Divorce may be influenced by such factors as loss of sexual attraction and emotional satisfaction and liberalization of divorce laws.
   3. Although divorce measures may be based on total population data, divorce-marriage ratios and final outcome data, each measurement technique is problematic.
   4. Trent and South's cross-national study assesses the impact of modernization, female labor force participation, the sex ratio and cultural belief systems on divorce.

B. Living Together, Marriage and Divorce
   1. Does living together prior to marriage reduce the number of potential unsatisfactory future unions, or does it signal a lack of commitment?
   2. Swedish and Centers for Disease Control studies indicate that divorce rates are high among couples that lived together prior to marriage.

C. Issues Confronting One-Parent Families
   1. One parent families may be result of divorce, death of a spouse or pregnancy of single women.
   2. Births among unmarried women have increased significantly since 1960.
   3. Single parents are more likely to face income concerns and increased demands on available time.

D. Studies Link Poor Parenting Skills and Childhood Deviance

E. Remarriage: Impact on Mate Selection, Children and Divorce
   1. Jacobs and Furstenberg study the association between mate selection and remarriage.
      a. Second husbands tend to have same social status as first husbands.
      b. Second husbands may have a higher or lower status than the first husband when a women is significantly older when she remarries.
      c. Second husbands often have a lower social status when a woman has younger children from a previous marriage.
   2. White and Booth study the association between remarriage and divorce and the impact of remarriage on stepchildren.

F. The "Empty Nest Syndrome" and Parental Emotional Readjustments

G. The "Postlaunch Honeymoon": White and Edwards
   1. Parents may experience an increase in marital happiness and life satisfaction when the last child leaves home.
   2. Parental happiness is contingent upon maintaining close contact with children that have left home.

H. Household Chores: Still Gender-Specific

Key Terms

Based on your reading of Chapter 13, you should be able to define and illustrate the key sociological concepts listed below. Page numbers are provided in parentheses as reference points.

- Kibbutz (367)
- Family (367)
- Marriage (368)
- Nuclear Family (368)
- Extended Family (368)
- Incest Taboo (369)
- Dependents (369)
- Final Outcome Measures (381)
- Cohabitation (384)
- Conjugal Careers (386)
- Empty Nest Syndrome (387)
- Postlaunch Honeymoon (388)

Key Research Studies

Listed below are key research studies cited in Chapter 13. Familiarize yourself with the major findings of these studies. Page references are provided in parenthesis.

Shorter: studied family ties in seventeenth century European extended families. Study challenges stereotypes concerning living conditions, parent-child ties, spouse ties and size of extended families (371-377).

Trent and South: conducted a cross-national study on factors impacting divorce. Strong associations exist between divorce and modernization, female labor-force participation and the sex ratio. Cultural belief systems (proportion Catholic) do not appear to impact divorce (383-384).

Living Together: Swedish study and Centers for Disease Control data indicate that divorce is high among persons living together before marriage (385).

Jacobs and Furstenberg: investigated mate selection patterns in remarriages. Second husbands and first husbands have similar social status unless a woman is older when she remarries or is bringing a young child from a previous marriage into the new household. In the first instance the second husband's social status may be higher or lower than the first husband's, but in the second instance, the second husband's social status is generally lower than the first husband's (386-387).

White and Booth: observed that the risk of divorce increases in remarriage if each partner has been previously divorced and if one or both partners bring a child from a previous marriage into the new household. Stepchildren tend to leave home at an earlier age (387).

White and Edwards: many parents experience an increase in marital happiness and life satisfaction once last child leaves home. However, parental satisfaction and happiness is contingent upon maintaining continued contact with children (388).

Info Trac Search Words

Enter these search terms to conduct more extensive investigations of key topics introduced in Chapter 13.

Extended Family

Browse through the studies located and select one that you find interesting. How does the description of the extended family in your chosen study compare with the picture portrayed by Shorter.

Dowry

Divorce

Subdivisions: Social Aspects

Select one of the studies cited. What outcome of divorce is being addressed? What are the major findings of your selected study. Are any recommendations offered?

Single-Parent Families

Remarriage and Children

Multiple Choice

Answers and page references are provided at the chapter end.

1. Each of the following components is included in Murdock's definition of the family except
   a. the family is a social group.
   b. the family is characterized by common residence.
   c. the family is characterized by economic cooperation.
   d. the family only needs to include one adult of either sex.
   e. the family is characterized by reproduction.

2. Reiss maintains that the key function of the family is
   a. economic cooperation.
   b. sexual gratification.
   c. socialization of children.
   d. regulating incest.
   e. all of the above

3. Each of the following statements is true for non-industrial societies except
   a. Polygamy is the rule.
   b. It is generally easier for a man to divorce his wife.
   c. Grooms are older than their brides.
   d. Couples eat meals together.
   e. The nuclear family is the norm.
4. Marriages generally

a. involve a formal commitment.
b. are intended to involve long-term relationships.
c. insure specific rights.
d. specify certain responsibilities.
e. all of the above

5. A family including a grandparent, grandparents or other adults from the parent's generation is known as a/an
   a. extended family.
   b. nuclear family.
   c. male headed household.
   d. female headed household.
   e. conjugal family.

6. Sexual relations between close-kin is generally prohibited. This is known as
   a. polygamy.
   b. monogamy.
   c. the dependent taboo.
   d. the incest taboo.
   e. conjugal marriage.

7. In modern societies pension plans, insurance and welfare programs assist families in meeting this need.
   a. emotional support
   b. sexual gratification
   c. economic support
   d. common residence.
   f. all of the above

8. Survey data reveal that persons residing in __________ are least likely to be satisfied with their home life.
   a. Japan
   b. France
   c. the United States
   d. Germany
   e. Mexico

9. During the seventeenth and eighteenth centuries, the average European extended family included
   a. two to three members.
   b. three to four members.
   c. five to six members.
   d. seven to eight members.
   e. ten or more members.

10. The small size of the preindustrial European family has been attributed to

a. high rates of infant and child mortality.
b. low rates of infant and child mortality.
c. high rates of infant mortality but low rates of child mortality.
d. low rates of infant mortality but high rates of child mortality.
e. increased life expectancy.

11. According to Shorter's research, which of the following statements about the seventeenth century European extended family is correct?
    a. Parent-child ties were strong.
    b. Spouse ties were strong.
    c. Adult peer group ties were strong.
    d. Neglect was uncommon with respect to child rearing.
    e. Homes were spacious and provided a great deal of privacy.

12. Which of the following statements about child care in the seventeenth century European extended family is incorrect?
    a. Medical records reveal that neglect was common.
    b. Instances of neglect were more likely to occur during harvest season.
    c. Infant deaths appear to have generated little regret or sorrow.
    d. Many infants were wet nursed.
    e. Each statement is true.

13. Modernization has been associated with
    a. an increase in the size of the family.
    b. continued neglect in child rearing.
    c. the maintaining of weak parent-child bonds.
    d. a strengthening of spouse ties.
    e. a strengthening of adult peer group ties.

14. In many modern societies, modernization has been associated with
    a. a decrease in the quantity of kinship bonds but continued stability in the quality of kinship bonds.
    b. an increase in the quantity of kinship bonds and an even greater increase in the quality of kinship bonds.
    c. a decrease in the quantity of kinship bonds and an even greater decrease in the quality of kinship bonds.
    d. an increase in the quantity of kinship bonds, but a substantial decrease in the quality of kinship bonds.
    e. the lack of change in either the quantity or quality of kinship bonds.

15. Among industrial nations, persons residing in __________ maintain the highest

levels of daily contact with their mothers.

a. Japan
b. Italy
c. Canada
d. Sweden
e. the United States

16. At any given moment __________ of North Americans report that they are currently divorced.

a. less than 5 %
b. 10 %
c. 25 %
d. 33 %
e. 50 %

17. Which of the following statements about divorce is false.

a. One-fourth of divorces occur between people who have children.
b. High divorce rates undoubtedly illustrate that marital relationships are much less important than they used to be.
c. Divorce has increased because divorce laws have become more restrictive.
d. Divorce rates can be easily measured in an accurate, valid manner.
e. Each of the statements is false.

18. Approximately __________ marriages end in divorce.

a. one in twenty
b. one in ten
c. one in five
d. one in three
e. one in two

19. *World Values Survey* data indicate that men and women residing in ________ are least likely to view extramarital affairs as being justified.

a. the United States
b. Nigeria
c. India
d. the Czech Republic
e. Mexico

20. Trent and South discovered that each of the following factors has a significant impact on divorce except

a. urbanization.
b. the proportion of the population Catholic.
c. female labor force participation.
d. economic development.
e. the sex ratio.

21. Which of the following Trent and South divorce study hypotheses is incorrectly stated.

a. As modernization increases, the divorce rate increases.
b. As women's labor force participation increases, the divorce rate increases.
c. As the sex ratio increases, the divorce rate decreases.
d. As the proportion of the population Catholic increases, the divorce rate decreases.
e. Each hypothesis is correctly stated.

22. A recent Centers for Disease Control study indicates that people who live together before eventually marrying
a. are less likely to divorce.
b. are more likely to divorce.
c. are no more likely to divorce than those who do not live together prior to marriage.
d. have more children.
e. have fewer children.

23. Currently __________ of all U.S. births are to unmarried women.
a. less than 2 %
b. 5 %
c. 10 %
d. 25 %
e. 33 %

24. Regardless of the structure of the family, research suggest that childhood deviance is often the result of
a. poor parenting skills.
b. male headed families.
c. female headed families.
d. extended families.
e. nuclear families.

25. The primary consequence of female-headed families is
a. steady employment.
b. divorce.
c. poverty.
d. high educational attainment.
e. adoption.

26. The recent Centers for Disease Control study indicates that __________ women are more likely to remarry within ten years of their divorce.
a. White
b. Asian
c. Native American
d. Hispanic
e. African American

27. When women with young children remarry, their second spouse is

a. more likely to be of a lower social status than the first husband.
b. likely to be of the same social status as the first husband.
c. more likely to be of a higher social status than the first husband.
d. more likely to have been divorced also.
e. less likely to have young children.

28. Divorces are higher in remarriages when
    a. only one of the partners has been divorced previously.
    b. both partners have been previously divorced.
    c. one or both partners bring children from a previous marriage into the new home.
    d. a and c
    e. b and c

29. White and Edwards have observed that parental marital happiness and life satisfaction increase once the
    a. first child leaves home.
    b. last child leaves home.
    c. the mortgage on the house is paid off.
    d. both couples retire.
    e. all their children have completed their education.

30. In many industrial societies today married men are least likely to perform which one of the following household chores?
    a. laundry
    b. small repairs
    c. grocery shopping
    d. cook
    e. care for the kids

True/False

Answers and page references are provided at the chapter end.

1. Two major themes dominating the sociological study of the family are the family is universal and the family is in a state of decline.

2. According to standard definitions of the family, the key function of the family is the provision of economic support for its members.

3. Nuclear families can be either two-generation or three-generation families.

4. Shorter's study suggests that most seventeenth century European extended families were quite large consisting of ten or more numbers.

5. Weak spouse ties and weak parent-child ties characterized the seventeenth century European extended family.

6. In the U.S. more than 75 percent of those who divorce remarry.

7. Survey data indicate that females are more likely than males to be involved in an extramarital affair.

8. A recent Swedish study indicates that persons who live together before marrying experience lower rates of divorce.

9. Research has linked childhood deviance with poor parenting skills.

10. A recent Centers for Disease Control study indicates that Conservative Protestant women are more likely to remarry within ten years of their divorce.

Short Answer Questions

These short answer questions are provided to test your knowledge and understanding of the basic sociological concepts presented in Chapter 13. Page references for answers are included at the chapter end.

1. Why is Murdock's definition of the family considered too restrictive by many sociologists?

2. Distinguish nuclear and extended families.

3. Identify and describe three traditional functions associated with the family.

4. What traditional extended family stereotypes are challenged in Shorter's study of seventeenth century European extended families?

5. How has modernization and industrialization strengthened spouse ties?

6. Identify several problems associated with calculating valid divorce rates?

7. According to the Trent and South cross-national study on divorce, what factors appear to impact the divorce rate?

8. Identify two specific issues impacting quality of life in one-parent families.

9. What factors tend to increase the risk of divorce in remarriages?

10. What is the "postlaunch honeymoon?"

Essay Questions

These questions are designed to test your understanding of key sociological concepts presented in Chapter 13 and your ability to apply these insights to concrete situations.

1. Identify five major social functions of families and indicate how these functions would vary in non-industrial and industrial societies.

2. Rural and urban lifestyles are significantly different. How would extended family networks vary in rural and urban environments? Would the extended family serve the same purpose and have the same functions in each environment?

3. Looking back at Chapter 12, how would the Guttentag and Secord sex ratios and sex roles theory help explain racial and ethnic variations in the U.S. divorce rate?

4. One-parent households have increased significantly since 1960. Do male headed households and female headed households share the same needs and concerns?

5. Social scientists are starting to reconsider the impact of birth order on human behavior and family socialization. How might the impact of divorce and remarriage on children vary by the child's birth order?

Answers

Multiple Choice

1. d (367)
2. c (367)
3. e (368)
4. e (368)
5. a (368)
6. d (369)
7. c (369-370)
8. a (371)
9. c (373)
10. a (372-373)
11. c (372-377)
12. e (374-375)
13. d (377-379)
14. a (379-380)
15. b (380)
16. a (380)
17. e (380-381)
18. e (381)
19. c (383-384)
20. b (383-384)
21. e (383)
22. b (384)
23. e (385)
24. a (385)
25. d (385)
26. a (386)
27. a (386-387)
28. e (387)
29. b (388)
30. a (388)

True/False

1. T (365)
2. F (367)
3. F (368)
4. F (372-373)
5. T (374-377)
6. T (380)
7. F (383)
8. F (384)
9. T (385)
10. T (386)

Short Answer

1. (367)
2. (368-369)
3. (369-371)
4. (372-377)
5. (378-379)
6. (381)
7. (383-384)
8. (385)
9. (386-388)
10. (388)

CHAPTER 14

RELIGION

Extended Abstract

Religion is a primary social institution. Humans have practiced religion for thousands of years. Religions are organized patterns of beliefs and practices addressing issues of ultimate concern and are grounded in a belief in the supernatural. It is the belief in the supernatural that separates religion from many different philosophies and political systems that may also address issues of ultimate concern. The supernatural is assumed in religious movements because humans cannot assure some needs, like afterlife. Religion can function to solidify the moral order and establish moral communities, but this function is contingent upon a religion's portrayal of God or Gods as conscious, powerful, morally concerned beings. Otherwise, there is no link between religion and morality. Also, Miller and Huffman argue that religious involvement is associated with risk-averse behavior rather than risk-taking behavior. Since men are more open to risk-taking behavior, this may account for the gender difference in religious involvement that favors women. However, this gender difference also appears to be grounded in the perception of the supernatural as a morally concerned being.

Stark maintains that among more complex societies, religious pluralism is common. The set of competing religious groups within any geographic area is known as the religious economy. The religious economy may be regulated or based on a deregulated, free-market. Low degrees of religious pluralism and religious involvement are associated with highly regulated religious economies. Religious markets may be regulated by the state or by strong religious monopolies.

Church-sect theory is introduced next to help explain the growth of religious pluralism in deregulated religious economies. Churches are more intellectually based organizations and are grounded in cosmopolitan networks; whereas, sects are more emotionally based organizations grounded in local networks. Niebuhr argued that churches would appeal to the middle and upper classes while sects would appeal to the lower class. Thus, religious denominational conflicts are expressions of class conflict. Niebuhr also proposed that churches appeal to the more affluent because they maintain a "this world" focus. Sects appeal to the masses because of their "otherworldly" focus. Needs that cannot be met in "this world" can be provided by the supernatural in the "next world." According to church-sect theory, new religious movements begin as sects. Over time, successful sects become churches. Since church organizations do not address the needs of the masses, new sects are created. This cycle of church-sect formation generates religious pluralism. Revising Niebuhr's church-sect theory, Johnson maintains that churches reflect and reinforce society's norms, while sects often challenge society's norms. Churches are low tension organizations while sects are high tension organizations.

Since church-sect theory suggests that some religious organizations are in the process of becoming more "this worldly" and also legitimate society's norms, some social scientists have argued that secularization is inevitable. The power and influence of religious organizations erodes as other organizations, like science, meet the needs that religious groups once satisfied. On the other hand, Stark maintains that secularization stimulates changes in the religious economy as revival movements (sects) and new religious movements (cults) emerge to meet important needs not addressed by older, less effective church-type groups. New religious movements and religious reform movements emerge where older faiths are weak, and secularization signals religious change rather than a decline in the influence of religion.

The religious economies of the United States, Canada, Europe and Latin America are described throughout the reminder of the chapter. The U.S. religious economy is characterized by a high degree of religious pluralism and religious involvement. Over 1,500 different religious denominations exist and almost 40 percent of U.S. residents attend religious services in any given week. Church-type groups, like the Christian Church (Disciples), are experiencing membership decline, but sect-type groups, like the Church of God in Christ, have experienced impressive gains. Likewise, cults are more prolific where established groups (church-sect types) are weak, and where a significant number of persons grow up in irreligious homes. In the U.S. this is particularly evident on the West Coast (the "Unchurched Belt") where church membership is low and cult involvement is high.

The Canadian religious economy is less diverse. Two out of every five Canadians are affiliated with the Catholic Church, and although a small percentage of the population is involved in cults, these movements are strong (Western cities) where traditional groups are weak. Research also indicates that European involvement in cults is more pronounced than previously assumed. The rate of cult movement formation in Europe is double the U.S. rate. Given the low levels of involvement in traditional religious organizations in certain areas of Europe, this finding is consistent with the predictions of the church-sect and religious economy theories.

The church-sect and religious economy arguments also provide a plausible explanation for the rapid growth of Protestantism in Latin America and the proliferation of religious involvement in Eastern Europe following collapse of the Soviet Union. The growth of Protestantism in Latin America may be perceived primarily as a response to the deregulation of the Catholic monopoly as many Latin American governments have been more tolerant of non-Catholic groups in recent decades. As more religious options are made available, religious involvement is stimulated. Likewise, as a result of the recent collapse of communist regimens in Eastern Europe, strong government opposition to religion has ceased. This has led to the creation of a deregulated religious economy. The revival of religious involvement in Eastern Europe is thus as expected.

Key Learning Objectives

After a thorough reading of Chapter 14, you should be able to:

1. identify the distinctive features of religion as a social institution.

2. clarify the association between religion and morality.

3. review explanations accounting for the gender differences in religious involvement.

4. specify the association between religious involvement and risk-averse behavior.

5. state the nature of the relationship between religious economies and religious pluralism.

6. critically evaluate Niebuhr's church-sect theory.

7. explain why churches are low tension organizations and sects are high tension organizations.

8. identify key tenets of secularization theory.

9. explain why revival (sects) and religious innovation (cults) are two outcomes of secularization.

10. distinguish sects and cults as two types of religious organization.

11. explain how charisma is related to the revival of existing religious traditions and the creation of new religious movements.

12. outline the distinguishing features of the U.S. and Canadian religious economies.

13. locate the "Unchurched Belt" and explain why it exists.

14. specify the conditions favoring the growth of cult movements.

15. apply the religious economy argument and the church-sect theory to cult development in Europe, the growth of Protestantism in Latin America and the revival of interest in religion in eastern Europe.

Chapter Outline

I. Religion as a Social Institution
   A. Religion and Neanderthal Ancestors
   B. Problems in Defining Religion
      1. Religion is a set of beliefs and practices.
      2. Religion attempts to answer questions of ultimate concern.
      3. The distinctive feature of religion is the belief in the supernatural.
   C. Religion and the Moral Order
      1. Religion supports the moral order when the religion is grounded in a perception of God or Gods as conscious, powerful, morally concerned beings.
      2. Religions grounded in the belief in impersonal essences are unable to solidify the moral order.
      3. The association between religious participation and morality is weak.
   D. The Gender Difference in Religious Involvement
      1. Thompson links the gender difference in religious involvement with personality differences. Persons scoring higher on a standardized femininity measure report higher levels of religious involvement.
      2. Miller and Hoffman portray religion as risk-averse behavior. Males are more likely to be risk-takers and thus avoid religious involvement.

3. Miller and Stark note that the gender difference in religious involvement disappears when the risks of being irreligious are low.

II. Religious Economies and Church-Sect Theory

A. Defining the Religious Economy

1. The religious economy is the set of competing faiths within a particular geographic region.
2. Religious pluralism is more pronounced in deregulated, free-market environments.

B. Church-Sect Theory: Niebuhr

1. Church-type organizations are intellectually based and grounded in cosmopolitan networks; whereas, sects are emotionally based and grounded in local networks.
2. Churches focus on "this world" orientations and attract the affluent while sects focus on "otherworldly" orientations and attract the poor.
3. Religious Pluralism is an outcome of the church-sect cycle. Successful sects eventually shift their interest to "this world" and become church-type organizations. Sects arise to challenge church-type organizations and serve the needs of the poor.
4. Churches reinforce society's norms and are low tension organizations, but sects challenge society's norms and are high tension organizations.

C. Secularization and Religious Change

1. Dissatisfaction with church-type religions stimulates religious change rather than a decline in religion.
2. Secularization promotes religious revival (sect formation) and religious innovation (cult formation).
3. New religious movements begin as cult movements, and new religious movements grow in environments where traditional religions are weak.
4. New religious movements are linked to dynamic, charismatic leaders and their ability to establish strong networks.

III. Comparing Different Religious Economies

A. The U.S. Religious Economy

1. The U.S. religious economy is characterized by a high degree of religious diversity and religious involvement.
2. Church-type denominations, like the United Church of Christ, are losing members, but sect type denominations, such as the Church of God, are gaining members.
3. Cults are strong on the West Coast (the "Unchurched Belt") where traditional religious groups (churches and sects) are weak.
4. Rapid population change is a major cause of low church membership rates in the western states.
5. Persons from irreligious backgrounds tend to join cults.

B. The Canadian Religious Economy

1. Compared to the U.S. religious economy, the Canadian religious economy is less diverse.
2. Two out of five Canadians are affiliated with the Catholic Church.
3. Cults are stronger in western Canadian cities where traditional religious groups are weak.

C Cult Development in Europe

1. Weekly church attendance is low in many European societies.
2. Cults are more pronounced in countries where conventional religious groups are weak.
3. The rate of cult movement formation in Europe is double the U.S. rate.

D. Latin America: A Deregulated Market

1. Catholic monopoly previously restricted alternative religious involvement.
2. The decline in government intolerance of non-Catholic religious groups resulted in the deregulation of the Latin American religious economy.
3. Protestantism flourishes in a deregulated market.

E. Religious Revival in Eastern Europe

1. Communist regimes promoted scientific atheism.
2. The collapse of communist regimes signaled the creation of a deregulated religious economy.
3. The current religious revival in Eastern Europe is consistent with a deregulated religious economy.

F. Otherworldly Religions Attract the Rich and the Poor

Key Terms

Based on your reading of Chapter 14, you should be able to define and illustrate the key sociological concepts listed below. Page numbers are provided in parentheses as reference points.

Religious Economy (393)
Secularization (393)
Revival (393)
Religious Innovation (394)
Questions About Ultimate Meaning (394)
Supernatural (395)
Religion (395)
Conscious Supernatural Beings (398)
Supernatural Essence (398)
Pascal's Wager (400)
Religious Pluralism (401)
Churches (401)
Sects (401)
Sect Formation (403)
Church-Sect Theory (403)
Cultural Innovation (405)
Cults (405)
Cult Formation (406)
Charisma (406)
Unchurched Belt (408)
Bible Belt (408)

Key Research Studies

Listed below are key research studies cited in Chapter 14. Familiarize yourself with the major findings of these studies. Page references are provided in parentheses.

Stark: Cross-national study reveals that religion does not always solidify the moral order. The link between religion and morality is contingent upon the perception of the supernatural as a conscious, powerful, morally concerned being or beings (396-399).

Thompson: links the gender difference in religious involvement to personality characteristics. Men and women scoring high on a femininity index are more likely to be religious (399).

Miller and Hoffman: associate religious involvement with risk-averse behavior. Persons scoring high on risk-taking behavior tend to be less religious. Men traditionally have been associated more with risk-taking behavior (399-400).

Miller and Stark: the gender difference in religious involvement disappears when the risk of being irreligious is low (400).

Niebuhr: developed the church-sect theory as a means of explaining denominational diversity. Churches are grounded in cosmopolitan networks that develop a "this world" focus and appeal to the more affluent. Sects are grounded in local networks that develop an "otherworldly" focus and appeal to the poor (401-403)

Johnson: revised the church-sect theory. Churches affirm society's norms and are low tension organizations while sects challenge society's norms and are high tension organizations (403).

Martin: documented the recent growth of Protestantism in Latin America. The rapid growth of Protestantism is associated with a deregulated religious economy (412-413).

Filatov and Furman: obtained access to government surveys from the former Soviet Union. Analysis of available data documents a decline in atheism in major Russian cities following the collapse of the former Soviet Union. Nontraditional religious beliefs were apparently common in the former Soviet Union even though religion was suppressed (414).

Info Trac Search Words

Enter these search terms to conduct more extensive investigations of key topics introduced in Chapter 14.

Religion and Sociology

Subdivisions: Analysis

Select one of the studies cited. See if you can determine the author's theoretical perspective. Specify the study's main findings.

Religious Involvement

Secularization

Review the articles listed and select one. Does the study provide evidence to support or challenge the claims of secularization?

Cults

Subdivisions: Social Aspects

Protestantism and Latin America

Multiple Choice

Answers and page references are provided at the chapter end.

1. The set of competing faiths within a society is known as
   a. secularization.
   b. the religious economy.
   c. the supernatural.
   d. cult formation.
   e. the "churched" belt.

2. Gods, morally concerned beings and impersonal forces are all examples of
   a. religion as a social institution.
   b. religion as a set of practices.
   c. religious market regulation.
   d. the supernatural.
   e. secularization.

3. Most people responding to surveys on religion state that they are religious. However, approximately 38% of persons residing in __________ indicate that they are atheists.
   a. the United States
   b. Russia
   c. China
   d. Japan
   e. France

4. Durkheim believed that religion united people into
   a. moral communities.
   b. supernatural states.
   c. religious economies.
   d. secular societies.
   e. atheistic societies.

5. Recent studies indicate that moral attitudes are primarily influenced by
   a. church attendance.
   b. mosque attendance.
   c. ritual behavior.
   d. belief in the God or the Gods as conscious supernatural beings.
   e. belief in the supernatural as an impersonal essence.

6. Miller and Hoffman have observed that religious involvement is higher among persons who are identified as
   a. risk averse.
   b. risk takers.
   c. exhibiting strong feminine qualities.
   d. exhibiting gender neutral qualities.
   e. exhibiting strong masculine qualities.

7. The coercion of religion by the state is an example of
   a. a regulated religious economy.
   b. a deregulated religious economy.
   c. a free-market religious economy.
   d. market demand.
   e. increased market choice.

8. Niebuhr developed the church-sect theory as a means of providing an explanation for
   a. the existence of the supernatural.
   b. gender differences in religious involvement.
   c. the emergence of state controlled religion.
   d. the high degree of denominational diversity associated with Christianity.
   e. none of the above

9. Church-type religious organizations
   a. stress intellectualism.
   b. stress emotionalism.
   c. are grounded in cosmopolitan networks.
   d. a and c
   e. b and c

10. Niebuhr argued that sects
    a. are grounded in local networks.
    b. appeal to the poor.
    c. stress the importance of the spiritual world.
    d. tend to maintain an otherworldly focus.
    e. each of the above

11. According to church-sect theory, religions originate as __________ address the needs of the __________.
    a. sects; lower class
    b. sects; upper class
    c. churches; upper class
    d. churches; lower class
    e. churches and sects; middle class

12. Churches tend to
    a. reinforce society's norms and are high tension organizations.
    b. reinforce society's norms and are low tension organizations.
    c. challenge society's norms and are low tension organizations.
    d. challenge society's norms and are high tension organizations.
    e. none of the above

13. Auguste Comte maintained that
    a. religion is an infantile illusion.
    b. religion is the opiate of the people.
    c. religion would eventually replace science.
    d. sociology is a scientific substitute for religion.
    e. sociology will eventually be replaced by religion.

14. According to Stark, secularization
    a. signals a decline in the influence of religion in daily life.
    b. promotes religious revival.
    c. stimulates the development of new religious movements.
    d. a and b
    e. b and c

15. Religious movements representing new religious traditions within a society are often labeled
    a. churches.
    b. sects.
    c. cults.
    d. religious economies.
    e. religious monopolies.

16. Which of the following statements is false?
    a. Cults generate a low degree of tension with the surrounding society.
    b. Cults are generally considered deviant religious groups.
    c. Religions begin as cult movements.
    d. Most new religions or cults fail.
    e. Each statement is true.

17. The ability of a person to inspire others and encourage people to develop strong attachments with others is know as
    a. moral attraction.
    b. religious attraction.
    c. charisma.
    d. anomie.
    e. secularization.

18. The largest religious group in the U.S. religious economy is the
    a. Southern Baptist Convention.
    b. Church of God in Christ.
    c. National Baptist Convention of the United States.
    d. Roman Catholic Church.
    e. Episcopal Church.

19. Approximately __________ of U.S. residents attend religious services.
    a. 5 %
    b. 20 %
    c. 33 %
    d. 40 %
    f. 60 %

20. Each of the following U.S. religious groups has experienced substantial membership declines over the 1960-2000 period except
    a. Unitarian-Universalists.
    b. Jehovah's Witnesses.
    c. Episcopal Church.
    d. Presbyterian Church (U.S.A.).
    e. United Methodist Church.

21. The "Bible Belt" refers to the belief that U.S. religious involvement is particularly strong in
    a. Alaska and Hawaii.
    b. the West.
    c. the Great Lakes region.
    d. the Northeast.
    e. the South.

22. In the "Unchurched Belt," church membership
    a. and cult membership are the same.
    b. is low and cult membership is high.
    c. is high and cult membership is low.
    d. and cult membership are high.
    e. and cult membership are low.

23. According to Stark, a major factor influencing low church membership rates on the West Coast is
   a. constant and rapid population change.
   b. increasing affluence.
   c. the presence of a large young population.
   d. high levels of educational attainment.
   e. low unemployment rates.

24. Most people joining cults
   a. come from religious backgrounds.
   b. are from the "Bible Belt.".
   c. grew up in the Catholic Northeast.
   d. are from a lower social class background.
   e. do not come from a religious background.

25. The largest religious group in the Canadian religious economy is the
   a. United Church of Canada.
   b. Church of Scientology.
   c New Age Movement.
   d. Roman Catholic Church.
   e. Anglican Church of Canada.

26. Cult groups in Canada tend to
   a. recruit more female members.
   b. recruit more male members.
   c. involve a small number of people from the total population.
   d. a and c
   e. b and c

27. Available research suggests that in Europe traditional church membership rates
   a. and cult membership rates are equal.
   b. are high and cult membership rates are low.
   c. are low and cult membership rates are high.
   d. and cult membership rates are very low.
   e. and cult membership rates are very high.

28. Cult movements appear to be most pronounced in this European nation.
   a. Switzerland
   b. Sweden
   c. Ireland
   d. Italy
   e. France

29. Which of the following statements about the Latin American religious economy is correct.
    a. The Latin American religious economy is becoming less diverse.
    b. The Catholic religious monopoly is becoming stronger.
    c. The government persecution of non-Catholic groups is increasing.
    d. Increasing religious pluralism has resulted in lower general rates of religious participation throughout Latin America.
    e. Membership in Protestant denominations is increasing.

30. Prior to the collapse of the former Soviet Union
    a. scientific atheism was part of the required educational curriculum in school.
    b. many churches, synagogues and mosques were closed, destroyed or used for other purposes.
    c. a few religious groups like the Russian Orthodox Church were placed under strict state control.
    d. religious leaders in Hungary were frequently arrested and charged with using religion to corrupt the morals of young people.
    e. each statement is true

True/False

Answers and page references are provided at the chapter end.

1. Archeologists have discovered evidence that religion has been practiced among humans for at least 100,000 years.

2. Sociologists have traditionally argued that all religions function to sustain the moral order.

3. Research indicates that persons who score high on masculinity measures are more religious.

4. Secularization is the set of competing religious groups within a given geographic area.

5. Church is to sect as cosmopolitan networks are to local networks.

6. Niebuhr maintained that sect-type organizations first appeal more to members of the upper and middle classes.

7. Two outcomes of secularization are sect revival and cult formation.

8. The "unchurched" belt refers to the low levels of church attendance characteristic of the Northeast.

9. Compared to the U.S., religious pluralism is more extensive in Canada.

10. Cult movement patterns in Europe provide evidence that supports the proposition that cults are strong where traditional religious groups are weak.

Short Answer Questions

These short answer questions are provided to test your knowledge and understanding of the basic sociological concepts presented in Chapter 14. Page references for answers are included at the chapter end.

1. Why is the belief in the supernatural a key element in providing an adequate sociological definition of religion?

2. Review recent research on the gender difference in religious involvement and state the findings.

3. What is the nature of the association between religious economies and religious pluralism?

4. Outline the major tenets of the church-sect theory.

5. How do churches, sects and cults differ?

6. Stark claims that secularization is an aspect of religious change. Identify two religious structural outcomes of secularization.

7. What is charisma, and how is it related to the formation and development of religious groups?

8. Distinguish the "Bible Belt" and the "Unchurched Belt."

9. Compare and contrast the U.S. and Canadian religious economies.

10. Identify major factors associated with the growth of Protestantism in Latin America.

Essay Questions

These questions are designed to test your understanding of key sociological concepts presented in Chapter 14 and your ability to apply these insights to concrete situations.

1. Explain how the social functions of religion would vary in non-industrial and industrial societies.

2. How does the perception of the supernatural regulate the association between religion and morality?

3. Critically evaluate secularization theory. Identify factors that support and challenge the claims of secularization theorists.

4. To what extent is the "Bible Belt" a regulated religious economy that is becoming more deregulated?

5. Recently the U.S. has experienced a significant increase in immigration from Asian countries and Latin America. How will this impact the U.S. religious economy over the next ten to fifteen years?

Answers

Multiple Choice

1. b (393)
2. d (394-395, 398)
3. c (396)
4. a (396)
5. d (398-399)
6. a (399-400)
7. a (400)
8. d (401)
9. d (401)
10. e (401-403)
11. a (402-403)
12. b (402-403)
13. d (404)
14. e (404-406)
15. c (405)
16. a (405-406)
17. c (406)
18. d (406)
19. d (406)
20. b (407)
21. e (408)
22. b (408-409)
23. a (408-409)
24. e (410)
25. d (411)
26. d (411)
27. c (412)
28. a (412)
29. e (412-413)
30. e (414-415)

True/False

1. T (393)
2. T (396)
3. F (399)
4. F (393, 400)
5. T (401)
6. F (402-403)
7. T (406)
8. F (408-409)
9. F (411)
10. T (412)

Short Answer

1. (394-395)
2. (399-400)
3. (400-401)
4. (401-403)
5. (401-403, 405-406)
6. (404-406)
7. (406)
8. (408-409)
9. (406-412)
10. (412-413)

CHAPTER 15

POLITICS AND THE STATE

Extended Abstract

This chapter addresses the provision of public goods, the development and function of the state and public opinion. The "tragedy of the commons" and the "free rider" phenomenon are cited as examples illustrating the problems associated with the provision of public goods. The common grazing area worked in medieval England as long as population growth was low and individual herders did not significantly increase herd size and overgraze the land. Initially, individual herders were able to increase their herd size at the expense of all herders. However, as increasing population growth occurred and each herder attempted to maximize individual benefits at the expense of the other herders, overgrazing occurred. Landowners attempted to protect the land by erecting fences in the common area. This restricted the freedom of individual herders and concentrated power among an elite group, landowners. Standard of living among herders was now stratified on the basis of herders who were able to obtain grazing rights. The "tragedy of the commons" demonstrates that the provision of collective goods often requires the sacrifice of individual power. Power is transferred to leaders or other political structures like the state.

The "tragedy of the commons" is reflected also in the "harvesting game" experiment of Messick and Wilke. Students were supposedly placed in three harvest level groups, overuse, underuse and optimal use. Within each group inequality was believed to be present. In the second part of the experiment, the participants of each group were informed that the majority of participants had agreed to elect leaders to represent them. Prior to this election, participants in each group increased their harvests over time. The increase was greater when participants believed they were being influenced by higher degrees of inequality. Once leaders (regulators) were elected, harvest size was reduced, harvests were equitably distributed among the other participants but the leaders took larger shares. Here again individuals will sacrifice power in order to receive a public good, and leaders will use their power to exploit others and maximize self-interest.

The provision of public goods is also associated with the "free rider" phenomenon. "Free riders" are those who benefit from a good, but do not share in its cost. The state arises to insure that all members will share in the cost associated with providing a public good. To obtain compliance the state is authorized to use force or coercion; however, representatives of the state may use their power to maximize self-interest.

The state, therefore, functions to enforce the norms and make life more predictable, to provide public goods and to insure protection from internal and external attack. Data from the *Standard Cross-Cultural Sample* reveal that the development of the state is linked to higher levels of agricultural development. However, warfare and a higher degree of stratification are also linked to state development. Thus, the potential for the abuse of power resides in the creation of the state. To control the state, eighteenth century political thinkers argued that rules must be established to regulate the state's power and power must be dispersed among groups with diverse interests. The signing of the Magna Carta by King John and the creation of the House of Lords and House of Commons are cited as early English attempts to employ these principles to control the state. In the U.S. state control is maintained by a system of checks and balances that integrate the executive, legislative and judicial branches of the federal government.

The English and U.S. examples of state control illustrate the pluralist state. In a pluralist state, many groups hold power. These minority power groups form coalitions and engage in conflict to maximize self-interest. The pluralist state differs from the elitist state, where power is concentrated in a single group, such as a political party. Stark maintains that democracy is sustained in the pluralist state. However, Mills notes that even within a pluralist state, a power elite exists. This power elite generally includes government leaders, business leaders and the military. Persons representing these areas tend to possess similar backgrounds and similar interests.

The remainder of the chapter is devoted to a discussion of political behavior and public opinion. The study of political behavior and public opinion in the U.S. has been heavily influenced by Gallup's work on opinion polling. The "*Gallup Poll*" was established in 1935 and has been used to gauge public opinion on key issues since that time. The various polls reveal that the media, elected officials and the general public do not always agree on their perception of political and social issues and that misperceptions and misrepresentations of public opinion are common.

U.S. election survey data indicate that U.S. residents do not fully exercise their right to vote and do not keep up with the daily news. Also, African Americans tend to favor Democrats while males, wealthier persons and persons with more education favor Republicans. *Gallup Poll* and *General Social Survey* data indicate that over time U.S. residents have been more willing to vote for a female candidate for president. However, few female candidates for public office have been elected at the national level. Does this mean that voters reject female candidates? Studies of voting patterns in Canada by Hunter and Denton and in the U.S. by Newman indicate that this is not the case. Few women have been elected because their odds for being elected are low. Since incumbents possess a significant political advantage and since the vast majority of incumbents are males, female candidates are less likely to be elected. Gender differences in electability disappear when the effects of party strength and incumbency are controlled.

Many social scientists assume that individuals are guided in their decision-making by their preference for particular world views or ideologies. Public opinion data reveal that this is not the case. Converse argues that as few as three percent of U.S. voters base their opinions on ideological preferences, and McClosky has shown that ideological preferences primarily impact the public opinion of members of small, elite groups. Panel studies, the polling of the same sample of respondents over time, indicate that voting patterns are impulsive and random and are not issue oriented. Political parties thus tend to develop broad public appeal rather than identifying with particular ideological positions. Consequently, political parties reflect similar positions on key issues.

Key Learning Objectives

After a thorough reading of Chapter 15, you should be able to:

1. provide examples of public goods.

2. explain the "tragedy of the commons."

3. indicate how collective and individual interests collide in the provision of public goods.

4. indicate why the "free rider" phenomenon is a problem encountered in the provision of public goods.

5. argue why leadership and the development of the state are central to the provision and maintenance of public goods.

6. identify the primary social functions of the state.

7. state the association among the development of the state and agricultural development, warfare and stratification based on data for non-industrial societies.

8. explain how the coercive power of the state is limited through laws and the dispersion of power.

9. distinguish elitist and pluralist states.

10. identify groups that may be included in the power elite and note how society may be controlled by the power elite.

11. note major trends in U.S. voting patterns.

12. outline Gallup's contributions to public opinion polling.

13. identify major factors impacting the electability of female candidates.

14. specify the association among voting patterns, ideology and party identity.

15. define panel studies and explain how they are useful tool for studying public opinion.

Chapter Outline

I. The Provision of Public Goods
    A. The "Tragedy of the Commons"
        1. Initially herders sharing common land can improve their position and incur only part of the cost.
        2. Population growth and increased herd size produce overgrazing.
        3. Landowners erect fences and limit grazing rights to maintain control over the land.
        4. The provision of a public good eventually requires the concentration of power in the hands of a few.
    B. A "Tragedy of the Commons" Experiment: Messick and Wilkie
        1. Students are divided into different harvest groups based on productivity.

2. Perceived inequality is experienced in each group, and leaders are elected.

3. Harvests increase in each productivity group over time prior to the election of a leader.
4. Leaders decrease harvest size, distribute proceeds equally among members and exploit group members by taking larger harvest shares.
5. Experiment shows how people utilize power to exploit others.

C. The "Free Rider" Phenomenon
1. As the size of a group increases, more people may enjoy the public goods provided by the group without having to bear the costs.
2. Coleman's classic example of multiple-person exchanges illustrates the "free rider" phenomenon.
3. States may use force and coercion to force individuals to share in the support of public goods.

II. The Development and Function of the State

A. The State as a Social Institution
1. The state exists to enforce norms, make life more predictable, provide public goods and insure protection from internal and external threats.
2. The state is granted the right to use force to gain compliance.

B. The Rise of the State in Non-Industrial Societies
1. In hunting and gathering societies, authority and power are linked to kinship and age.
2. The rise of the state is correlated with increased agricultural development, increased warfare and increased stratification.

C. Controlling the State's Power
1. The state's power is specified and limited through the creation of laws.
2. The power of the state is limited when power is divided among diverse groups.
   a. The Magna Carta, the House of Lords and the House of Commons represent early English attempts to control the power of the state.
   b. In the U.S. power is distributed among the executive, legislative and judicial branches of government.

D. Elitist States, Pluralist States and the Power Elite
1. In an elitist state, power is concentrated in one ruling group such as a one-party state.
2. Democratic states tend to be pluralist states, and power is distributed among diverse groups. These groups may form coalitions and engage in conflict as each group maximizes its self-interest.
3. Mills argues that government leaders, business leaders and the military are part of the power elite. Members of these groups share similar backgrounds and interests.

III. Public Opinion and Political Behavior

A. Public Opinion Polling: Gallup

1. Gallup established the American Institute of Public Opinion to gauge public opinion on important political, social and moral issues.
2. Gallup Poll findings often reveal that the media, elected officials and the general public do not always share the same view on key issues.
3. Misperception and misrepresentation of public opinion are common.

B. U.S. Voting Patterns

1. Polls indicate that U.S. voters have little interest in the political process.
2. Men, more affluent persons and persons with a higher degree of educational attainment are more likely to support Republican candidates.

C. Voting for Female Candidates: U.S. and Canadian Case Studies

1. Poll suggest that voters are willing to support female candidates, but few female candidates are elected.
2. Political incumbents maintain a significant advantage.
3. Most incumbents are male.
4. Gender differences in voting outcomes disappear when controls for party strength and incumbency are introduced.

D. Ideology and Voting Patterns

1. Many social scientists have assumed that ideological perspectives and world views impact individual voting patterns and opinions.
2. Individual responses to public issues are often based on impulse and are random.
3. Panel studies involve re-interviewing the same sample over time and are used to evaluate consistency in voter response patterns.
4. Members of elite groups tend to be influenced more by ideology.
5. Due to the weak link between ideology and individual voting patterns, political parties develop broad-base appeal and reflect similar positions on key issues.

Key Terms

Based on your reading of Chapter 15, you should be able to define and illustrate the key sociological concepts listed below. Page numbers are provided in parentheses as reference points.

The Tragedy of the Commons (420)
Collective Goods and Public Goods (421; 423)
State (424)
Free Riders (425)
Pluralism (429)
Tyranny of the Minority (430)
Tyranny of the Majority (430)
System of Checks and Balances (430)
Elitist State (430)
Pluralist State (430)
Power Elite (432)
American Institute of Public Opinion / *Gallup Poll* (434-435)
World Views (440)
Meaning Systems (440)
Ideology (440)
Panel Studies (442-443)
Issue Publics (443)

Key Research Studies

Listed below are key research studies cited in Chapter 15. Familiarize yourself with the major findings of these studies. Page references are provided in parentheses.

Lloyd: commented on the "freedom of the commons" and the "tragedy of the commons." Overgrazing of the commons led to the establishing of grazing rights, which limited access to the land. An inherent problem associated with the provision of public goods is the loss of individual power (419-420).

Messick and Wilke: illustrated the "tragedy of the commons" in a controlled experiment. When authority for controlling harvesting is delegated to leaders, leaders decrease harvest size and distribute harvest shares equally among participants but keep a larger share for themselves. Study illustrates that people will use power to promote self-interest (421-423).

Olsen: force must be employed to assure that individuals will share the costs associated with the provision of public goods (423-424).

Coleman: illustrated the "free rider" phenomenon among multiple-person exchanges. Individuals are less likely to pay the costs associated with providing public goods as the size of the group increases (425).

Mills: developed concept of the power elite. Power within a society is concentrated among political leaders, business leaders and the military. Members of the power elite share similar backgrounds and interests (432).

Gallup: founder of modern opinion polling. Public opinion polls frequently illustrate a significant degree of divergence on major issues among the media, public officials and the general public (434-435).

Hunter and Denton: studied Canadian voting patterns for female candidates and linked female candidates' electoral success to opportunities for being elected. Weak party influence and incumbent advantage limit female candidates' odds of being elected (437-438).

Newman: studied U.S. voting patterns for female candidates. Data indicate that incumbents hold an advantage, and the majority of incumbents are males (438-439).

Converse: ideology has little impact on public opinion. Voting patterns are influenced by impulse and vary randomly. Issue publics associated with any given topic involve only a small percentage of the general population (441, 443)

McClosky: ideology is most likely to influence the public opinion of members of elite groups, such as delegates to political conventions or highly-educated individuals (442).

Info Trac Search Words

Enter these search terms to conduct more extensive investigations of key topics introduced in Chapter 15.

Political Sociology

Tragedy of the Commons

Select one of the references listed and note how the "tragedy of commons" is being illustrated.

Power Elite

Select one of the studies cited and identify the groups included in the power elite. Does your selected study include the same groups Miller identified? How is power created and maintained by the power elite?

Gallup Poll

Ideology

Subdivisions: Social Aspects

Multiple Choice

Answers and page references are provided at the chapter end.

1. "Freedom of the Commons" was successful in medieval England only when the number of people on the estate was
   a. small and herd size was large.
   b. large and herd size was large.
   c. large and herd size was small.
   d. small and herd size was small.
   e. population size and herd size were equal.

2. The "tragedy of the commons" resulted in
   a. overgrazing.
   b. soil erosion.
   c. the erection of fences.
   d. the issuing of grazing rights to a few.
   e. all of the above

3. The elitist state generally leans toward __________ and the pluralist state tends to allow for more __________.
   a. individual freedom; tyranny
   b. tyranny; individual freedom
   c. public goods; private goods
   d. pluralism; monopolies
   e. collective freedom; individual freedom

4. The "harvesting" experiment indicates that
   a. all groups reduce their harvests over time.
   b. leaders use power to exploit others.
   c. leaders routinely increase harvest size to maximize output.
   d. leaders rewarded harder workers with more harvest shares.
   e. each condition was observed

5. When it comes to public goods
   a. individual self-interest is maximized by contributing to the "public good",
   b. individual self-interest is minimized by contributing to the public good.
   c. individuals must be forced to contribute their share of the cost of the public good.
   d. a and c
   e. b and c

6. Max Weber maintained that the state
   a. derives its power from religion.
   b. shares its power with the educational system.
   c. is able to claim a monopoly on the legitimate use of physical force.
   d. derives its power from the public ownership of the means of production.
   e. is an illusion of the oppressed.

7. In providing public goods, individuals are able to benefit from the good without having to share the costs associated with providing the good. This is known as the
   a. free rider phenomenon.
   b. power elite.
   c. the tyranny of the commons.
   d. elitist states.
   e. pluralist states.

8. The primary function of the state is to
   a. preserve internal order.
   b. make life more predictable.
   c. provide external security.
   d. provide collective goods.
   e. all of the above

9. Among non-industrial societies, the increased development of the state is associated with
   a. higher levels of agricultural development.
   b. more frequent warfare.
   c. less frequent warfare.
   d. a and b
   e. a and c

10. The increased development of the state is also associated with
    a. more frequent warfare.
    b. a low degree of stratification.
    c. a high degree of stratification.
    d. a and b
    e. a and c

11. The philosopher-king concept is primarily associated with
    a. Weber.
    b. Mills.
    c. Madison.
    d. Plato.
    e. Marx.

12. Two strategies for controlling the state's coercive power are
    a. the establishing of rules and laws.
    b. the concentration of power in one social group.
    c. the dispersion of power among several groups.
    d. a and b
    e. a and c

13. According to Stark English democracy is grounded in the right to
    a. bear arms.
    b. use coercion to deprive people of their property..
    c. public property.
    d. private property.
    e. taxation without representation.

14. The House of Lords and the House of Commons are examples of
    a. an elitist state.
    b. free riders.
    c. interest publics.
    d. political pluralism.
    e. freedom of the commons.

15. The use of representative government to exploit and abuse minorities is known as the
    a. tragedy of the Commons.
    b. tragedy of the Lords.
    c. free rider problem.
    d. tyranny of the minority.
    e. tyranny of the majority.

16. Each of the following is a characteristic of a pluralist state except
   a. the tendency toward dictatorship.
   b. the existence of many competing ruling elites.
   c. the dispersion of power among minority groups.
   d. a tendency toward democracy.
   e. more power struggles and coalitions.

17. According to Mills the power elite include
   a. business leaders, religious leaders and musicians.
   b. reporters, doctors and lawyers.
   c. business leaders, politicians and the military.
   d. ethnic minorities, women and the police.
   e. religious leaders, journalists and coaches.

18. Mills maintained that the power elite were primarily
   a. male.
   b. Protestant.
   c. from old East Coast families.
   d. educated at Ivy League Schools.
   e. all of the above

19. According to data from the *World Values Survey*, persons residing in __________ are most likely to view politics as important in their lives.
   a. United States
   b. South Korea
   c. Japan
   d. France
   e. India

20. The father of opinion polling is
   a. George Gallup.
   b. Karl Marx.
   c. C. Wright Mills
   d. Max Weber.
   e. Rodney Stark.

21. Which of the following statements is false.
   a. Gallup founded the American Institute of Public Opinion.
   b. The *Gallup Poll* collects data on political, social and moral issues.
   c. *Gallup Poll* data often reveal that the media and elected officials misrepresent public opinion.
   d. The *Gallup Poll* has affiliates in more than fifty nations.
   e. Each statement is true.

22. Based on the results of election surveys, each of the following statements is true except
    a. African Americans tend to vote Democratic.
    b. less educated persons tend to vote Republican.
    c. more affluent persons tend to vote Republican.
    d. men are slightly more likely to favor Republicans.
    e. each statement is true.

23. Canadian women received the right to vote
    a. before U.S. women received the right to vote.
    b. after U.S. women received the right to vote.
    c. before Canadian men received the right to vote.
    d. before U.S. men received the right to vote.
    f. before British women received the right to vote.

24. Hunter and Denton note that women experience problems being elected to office in Canada. Each problem is correctly stated except
    a. most incumbents are males.
    b. parties with losing records are more likely to nominate female candidates.
    c. women are disproportionately nominated for "lost cause" sects.
    d. Canadian voters are sexist and will not support female candidates.
    e. each problem is correctly stated

25. Newman's study of recent U.S. voting patterns reveals that
    a. U.S. voters prefer male candidates.
    b. U.S. voters prefer female candidates.
    c. male and female candidates are equally likely to win open seats.
    d. political action committees are more likely to fund female candidates.
    e. female incumbents are more likely than male incumbents to win re-election.

26. A theory about life based on a few general, abstract ideas is a/an
    a. democracy.
    b. hypothesis.
    c. ideology.
    d. panel study.
    e. collective good.

27. Converse has observed that about __________ percent of U.S. voters base their decisions on ideological beliefs they hold.
    a. 3
    b. 10
    c. 25
    d. 50
    e. 67

28. Which of the following statements is false.
    a. People rarely invent their own ideologies.
    b. A significant number of voters possess meaningful political beliefs.
    c. Many people respond to political issues based on impulse.
    d. Over time, many voters respond to issues in a meaningless, random manner.
    e. Elite groups tend to create and maintain ideologies.

29. In studies of public opinion, the same sample of voters is often interviewed at different points in time on the same and different issues. This type of study is know as a/an __________ study.
    a. event
    b. panel
    c. issue
    d. ideological
    e. incumbent

30. According to the *World Values Survey*, residents of _________ maintain the strongest ties to a particular political party.
    a. United States
    b. Japan
    c. Mexico
    d. Australia
    e. Sweden

True/False

Answers and page references are provided at the chapter end.

1. Fences and the need for grazing rights reflect the "tragedy of the commons."

2. In providing public goods, the number of free riders increases as the size of the group decreases.

3. Among non-industrial societies, political development increases as warfare increases.

4. A one-party government is normally associated with pluralist states.

5. According to Mills the power elite include religious leaders, labor union bosses and the police.

6. For many U.S. residents, the interest in politics is low.

7. The willingness of U.S. residents to vote for a female president has increased dramatically from 1937 to the present.

8. Social scientists have generally assumed that most individuals ascribe to a particular ideology or world view that shapes their views on key issues.

9. Ekstrand and Eckert observed that religious affiliation and church attendance exert a strong influence of peoples' willingness to vote for female candidates.

10. Political parties have learned that it is necessary to develop a broad-based appeal since ideology exerts a major influence on voter behavior.

Short Answer Questions

These short answer questions are provided to test your knowledge and understanding of the basic sociological concepts presented in Chapter 15. Page references for answers are included at the chapter end.

1. Distinguish the "freedom of the commons" and the "tragedy of the commons?"

2. Provide an example of a public good and indicate how the free rider problem may impact the chosen public good.

3. What is the major function of the state according to Weber?

4. Among non-industrial societies, what is the nature of the association among the development of the state and agricultural development, warfare and stratification?

5. Identify two social mechanisms designed to control the coercive power of the state.

6. How do "tyranny of the minority" and "tyranny of the majority" differ?

7. Clearly distinguish the elitist state and the pluralist state.

8. According to Mills, what groups are included in the power elite?

9. What has Gallup repeatedly discovered about the nature of public opinion?

10. What impact does of ideology have on public opinion?

Essay Questions

These questions are designed to test your understanding of key sociological concepts presented in Chapter 15 and your ability to apply these insights to concrete situations.

1. Apply the "tragedy of the commons" concept to the need to protect environmental resources like air, water and land.

2. Students in a class are required to participate in a group project that will result in a group grade. How might the instructor minimize the "free rider" effect?

3. Make a case for including the media as part of the power elite in contemporary U.S. society.

4. What are some of the potential sources of bias in public opinion polls?

5. Provide a critical response to the following statement, "Since public opinion is not influenced by ideology, political candidates do not need to develop political platforms."

Answers

Multiple Choice

1. d (419)
2. e (419-420)
3. b (421)
4. b (422-423)
5. e (423-424)
6. c (424)
7. a (425)
8. e (426)
9. d (427-428)
10. e (427-428)
11. d (428)
12. e (428)
13. d (428)
14. d (429)
15. e (430)
16. a (430-431)
17. c (432)
18. e (432)
19. b (433)
20. a (435)
21. e (434-435)
22. b (435)
23. a (437-438)
24. d (438)
25. c (439)
26. c (440)
27. a (441)
28. b (442-443)
29. b (443)
30. e (443)

True/False

1. T (419-420)
2. F (425)
3. T (428)
4. F (430)
5. F (432)
6. T (435)
7. T (438-439)
8. T (440)
9. F (441)
10. F (444)

Short Answer

1. (419-420)
2. (423-425)
3. (424)
4. (427-428)
5. (428)
6. (430)
7. (430-431)
8. (432)
9. (435)
10. (440-444)

CHAPTER 16

THE INTERPLAY BETWEEN EDUCATION AND OCCUPATION

Extended Abstract

The link between education and occupation is evident in hunter-gatherer societies and in industrial societies. Persons in these societies receive training, either at home, in the community or from schools, to prepare them for the tasks they will eventually perform. In industrial societies, a strong association exists among education, income and occupational prestige. Persons with higher education levels often are able to earn higher salaries from higher prestige occupations. Researchers have devised ways of ranking jobs on the basis of the prestige associated with the job. Consistency in prestige rankings has been demonstrated cross-nationally, and higher prestige jobs are associated with higher educational attainment levels. Educational training for particular types of occupations begins early as children are often placed in different educational tracks based on academic aptitude.

Technological changes have enabled workers in industrial societies to work smarter rather than harder. Taylor's time and motion studies are classic examples of how available technology and knowledge can be used to improve worker productivity. Taylor's approach is also known as "scientific management." Women's labor force participation increased substantially throughout the twentieth century. Factors associated with this increase include a low sex ratio, a decline in fertility, freedom from time consuming household tasks, a shift away from physically demanding jobs and the need for money. However, gender differences in income persist.

With a highly educated, highly trained workforce, persons with less education find that employment opportunities are limited. Unemployment data are based on persons of legal age who are not employed but are looking for a job. Unemployment is generally higher when jobs are in greater supply because more people are actively looking for work. When jobs are in short supply, married women and young people are less likely to look for work. Chronic unemployment is often associated with higher risks of poverty. Finally, U.S. workers tend to express high degrees of job satisfaction. Survey data indicate that most people would continue to work even if they were financially secure.

Stark next turns to a discussion of education. With the transition from an economy based on manual labor to one based on knowledge and technology, education has become even more important. Thus, what are some of the U.S. educational attainment trends, and how effective is the U.S. educational system? In 1647 the Puritans were among the first to insist upon mass education. An increasing demand for mass education was associated with the Industrial Revolution; however, the majority of U.S. children of school age did not complete high school until after WWII. With the present shift to an information-based economy, the demand for a college education and a professionally trained work force has increased. This has led some researchers to question the quality of U.S. education. Education is to be distinguished from schooling. Education addresses what a person learns; whereas, schooling refers to the number of years spent in school. In a study on U.S. literacy conducted by the Educational Testing Services

(ETS), Kirsch discovered that half the persons surveyed scored in the lower levels of each of the three literacy dimensions measured: quantitative, documentary and prose.

Findings such as these have led researchers to question the relevance and functions of schooling. Does schooling improve literacy, or does it reinforce the social and economic advantages or disadvantages associated with a child's home and neighborhood environment? In a mid-1960 study of educational opportunities and environments, Coleman concluded that there was little difference in the quality of schooling provided to different racial and ethnic groups and that quality of educational opportunities had little impact on educational achievement. However, a link was discovered between student performance and teacher competence. On the other hand, Heyns' study of the impact of the summer vacation months on the academic achievement of elementary and middle school children suggests that schooling does make a difference. Comparing student achievement test scores for periods when school was in session with scores at the end of the summer months, Heyns noted that scores generally did not increase as rapidly during the summer months. However, the decline in the rate of achievement was not as noticeable among students from more affluent families. Heyns discovered that the achievement differences could be attributed to the fact that the more affluent children were more likely to read more over the summer months. In a similar vein, Alexander, Natriello and Pillar tracked high school students who stayed in school and those that dropped out. Gains in educational achievement were less evident among dropouts.

Data from the "High School and Beyond Study" suggest that studying and academic performance are correlated. Asian American students are more likely to do homework and less likely to drop out. Compared to students in Catholic and private schools, students in public schools do less homework, are less likely to expect to go to college, are more likely to report witnessing behavior problems in school and are more likely to dropout of school. An increasing number of U.S. parents are choosing to homeschool their children. Educational professionals argue that homeschooling primarily benefits more affluent families, is inferior in quality and offers children limited socialization opportunities. However, a recent survey on homeschooling conducted by the National Center for Educational Statistics indicates that homeschooling is not primarily an option for more affluent families. Children who homeschool perform better on standardized tests, and households that homeschool do have one parent who either stays at home or works at home. Finally, cross-national studies on the effects of schooling indicate that students in more industrialized countries learn more from their educational experiences. However, the greatest advantages provided by an educated work force are observed among the poorer nations where social and economic background appears to have little effect on student performance.

The value and function of education is addressed in the final section. Again, the economic gains associated with education are most evident in less developed nations. In developed nations, education does not guarantee success, but education is necessary if one is to be successful. In the U.S. mean annual income increases as level of educational attainment increases. On the other hand, the value of college education has declined in industrial societies as the supply of college graduates has increased. This has led some researchers to suggest that the U.S. is becoming a "credential society." Allocation theories argue that educational credentials function to limit the number of persons who may enter certain occupations, thus making them less replaceable. Similarly, Meyer argues that educational attainment levels represent distinct social statuses that continue to impact a person's identity and quality of life throughout life. Educational institutions confer status and have the power to establish new knowledge classes or statuses by creating and approving new areas of technical expertise.

Key Learning Objectives

After a thorough reading of Chapter 16, you should be able to:

1. state the general association among educational attainment, income and occupational prestige.

2. describe how technological innovations have transformed the work place in industrial and industrializing societies.

3. outline the development of "scientific management."

4. identify important sociological factors associated with the increase in women's labor force participation.

5. indicate how unemployment is measured, and how the supply of jobs impacts the unemployment rate.

6. identify groups and areas that have experienced chronic unemployment.

7. summarize important trends in job satisfaction among U.S. workers.

8. briefly outline the rise of mass education in the U.S.

9. distinguish education and schooling and critically evaluate the quality of U.S. education.

10. assess the impact of summer vacation months on student achievement.

11. utilize available survey data to compare and contrast the effectiveness of public, private and Catholic schools.

12. note some of the major issues associated with the homeschooling phenomenon.

13. evaluate the association between educational attainment and success and determine whether education is a good investment.

14. demonstrate how educational systems utilize credentials to control worker replaceability.

15. describe how educational attainment levels function as distinct social statuses.

Chapter Outline

I. Occupational Prestige and the Transformation of the Work Place
   A. Occupational Prestige
      1. Higher educational attainment is often associated with high income and higher occupational status.
      2. Hatt and North developed occupational prestige ranking scale.
      3. Cross-national studies of occupational prestige indicate that jobs share similar prestige rankings.
   B. Educational Tracking and Early Occupational Socialization
   C. Technological Change and the Structural Transformation of Work
      1. People use technology to work smarter rather than harder.
      2. Taylor conducted time and motion studies as way of increasing worker efficiency (scientific management).
   D. Women's Labor Force Participation
      1. A low sex ratio has stimulated women's labor force participation.
      2. Fertility levels are down.
      3. Technological innovations have freed women from time consuming household tasks.
      4. Fewer new jobs rely on manual labor.
      5. More women are working for monetary reasons.
   E. Unemployment: Definition and Trends
      1. Only persons of legal age who are without a job and are actively looking for a job are included in the unemployment data.
      2. Unemployment rates are often high when jobs are more plentiful.
      3. Married women and young adults are less likely to be employed when jobs are in short supply.
      4. Poverty is a major problem associated with chronic unemployment.
      5. Unemployment varies significantly by region, age, race and ethnicity.
   F. Job Satisfaction Trends
      1. Survey data reveal that U.S. workers express high levels of job satisfaction.
      2. The majority of U.S. workers indicate that they would continue to work even if they were financially secure.

II. Education in Industrial Societies
   A. The Growth of Mass Education
      1. Mass education begins in the U.S. with the Puritans.
      2. Mass education is linked to the Industrial Revolution.
      3. The majority of U.S. residents did not begin completing high school until after WWII.
      4. Today most U.S. and Canadian young adults enroll in college.
   B. Education Versus Schooling
      1. Education focuses on literacy and attainment while schooling deals with the time spent obtaining an education.

2. Educational Testing Service survey reveals that literacy levels among U.S. residents are low.
3. Coleman's study suggests that student achievement is not influenced by the quality of schools but is influenced by teacher competence.

4. Study by Heyns on variations in student achievement levels during the summer months indicates that schooling matters. Affluent children maintain higher achievement levels during summer months because they read more.
5. Study by Alexander, Natriello and Paller indicates that educational achievement is higher among persons staying in high school.

C. Trends in U.S. High School Education: The High School and Beyond Study
1. Asian Americans are more likely to study and less likely to drop out.
2. Dropout rates are higher in public schools than in Catholic or private schools.
3. Students in public schools are more likely to be exposed to behavior problems in school and are less likely to see themselves as going to college.

D. Homeschooling
1. The number of U.S. households homeschooling is steadily increasing.
2. Those who object to homeschooling argue that homeschooling favors affluent families, provides inferior education and limits children's socialization opportunities.
3. A recent national survey on homeschooling reveals that homeschooling does not favor affluent families. Househols that homeschool have one parent who stay at home or works from home, and homeschoolers tend to perform better on standardized tests.

E. Global Educational Trends: Heyneman and Loxley
1. Students in industrial nations tend to learn more while in school.
2. The economic benefits of education are demonstrated more clearly among poorer nations.
3. Family background has less impact on student achievement in poorer nations.

F. Education, Success and Status
1. Education is linked to success, but does not guarantee success.
2. The value of a college degree declines as the supply of college graduates increases.
3. Mean annual income in the U.S. varies significantly by educational attainment among adults aged 45-55.
4. According to allocation theories, educational credentials function to make workers less replaceable.
5. Meyer argues that educational levels represent distinct social statuses that impact a person throughout life. Educational institutions can define new knowledge areas and create new knowledge classes.

Key Terms

Based on your reading of Chapter 16, you should be able to define and illustrate the key sociological concepts listed below. Page numbers are provided in parentheses as reference points.

Occupational Prestige (450)
Educational Tracking (452)
Scientific Management (454)
Labor Force (454)
Unemployed (456)
Education (459)
Schooling (459)
Literate, Literacy (460)
"High School and Beyond Study" (463)
Statistical Abstract of the United States (468)
Credential Society (469)
Allocation Theories (469)

Key Research Studies

Listed below are key research studies cited in Chapter 16. Familiarize yourself with the major findings of these studies. Page references are provided in parentheses.

Hatt and North: developed an occupational prestige scoring and ranking system (450).

Porter and Pineo: studied occupational prestige in Canada. Discovered that Canadian and U.S. prestige rankings were highly correlated (450, 452).

Blau and Duncan: devised method for using average education and average income to predict occupational prestige score (452).

Taylor: conducted time and motion studies to evaluate worker efficiency. Taylor was a pioneer in "scientific management" (453-454).

Kirsch: directed study by Educational Testing Service (ETS) on U.S. literacy. Kirsch discovered that half of U.S. respondents scored low on the literacy dimensions measured (460-461).

Coleman and colleagues: observed an association between school quality and student achievement. Teacher competence influences student achievement (461-462).

Heyns: compared school year educational gains of elementary and middle school children with educational gains achieved over the summer months. More gains occurred during the school year, but children from more affluent backgrounds lost less ground over the summer months because of more extensive reading (462-463).

Alexander, Natriello and Pallas: conducted two-year study of high school students' cognitive aptitudes. Cognitive aptitudes were lower among dropouts compared to students who stayed in school (463).

High School and Beyond Study: large national survey of high school students sponsored by U.S. Department of Education. Selected findings indicate variations in study habits and dropout rates by race, ethnicity and type of school (463-465).

National Center for Educational Statistics: conducted national survey on homeschooling. Study indicates that homeschooling is not primarily limited to more affluent families. Households that homeschool have one parent who stays at home or works at home, and homeschoolers tend to perform better on standardized tests (466).

Heyneman and Loxley: conducted cross-national study of school effects. Discovered that children in industrialized nations learn more in their educational environments, but the economic advantages of education are more evident among poorer nations (466-467).

Collins: educational credentials function to control entrance into certain occupations. Credentials are used to manipulate the degree of replaceability associated with a task (468-469).

Meyer: developed a theory linking educational attainment and social status. Status identifications impact a person throughout life. Educational systems have the power to create and certify new areas of knowledge and establish new social statuses (469-471).

Info Trac Words

Enter these search terms to conduct more extensive investigations of key topics introduced in Chapter 16.

Social Status and Work

Job Satisfaction

Subdivisions: Analysis

Select one of the articles identified. How is job satisfaction being measured? How does job satisfaction vary? Identify the major findings of your selected study.

Knowledge Economy

Educational Sociology

Educational sociology is a sociological sub-field. Review the articles cited and select one that interests you. What sociological aspects of education are being addressed? See if you can identify the author's theoretical perspective.

Stratification and Education

Multiple Choice

Answers and page references are provided at the chapter end.

1. In Hunting and Gathering societies, education is primarily the responsibility of
   a. the religious leaders.
   b. the school system.
   c. teachers who periodically travel from village to village.
   d. the family.
   e. all of the above

2. Occupational prestige scores are generally based on
   a. subjective ratings of the value of a particular job.
   b. educational background of one's parents.
   c. the number of years of education the job requires.
   d. the salary associated with the job.
   e. the benefits associated with the job.

3. Each of the following is ranked as a high status occupation except
   a. physician.
   b. bartender.
   c. minister.
   d. college professor.
   e. building contractor.

4. Occupational socialization
   1. links education and occupation.
   b. begins early in life.
   c. is an example of differential socialization.
   d. involves educational tracking.
   e. all of the above

5. The main goal of "scientific management" is to
   a. control unemployment.
   b. increase wages.
   c. improve worker efficiency.
   d. increase the number of women in the labor force.
   e. offer equal opportunity and reduce discrimination in the labor force.

6. In the contemporary U.S. work place
   a. white-collar workers outnumber blue-collar workers.
   b. blue-collar workers outnumber white-collar workers.
   c. manual labor is increasingly being replaced by mental labor.
   d. a and c
   e. b and c

7. Each of the factors listed below have been associated with increased female labor force participation except
   a. the decline in the sex ratio.
   b. an increase in fertility.
   c. more freedom from the time consuming demands of housework.
   d. the need for more money.
   e. the increase in new jobs stressing mental rather than physical skills.

8. Among the nations listed below, women's labor force participation is highest for women living in
   a. the United States.
   b. Mexico.
   c. Japan.
   d. France.
   e. Iceland.

9. Presently women comprise approximately __________ percent of the total U.S. labor force.
   a. 25 %
   b. 33 %
   c. 47 %
   d. 55 %
   e. 67 %

10. U.S. women are least likely to be employed in this profession.
   a. engineer
   b. religious leader
   c. dentist
   d. college professor
   e. author

11. The term, unemployment, is applied
   a. to all persons looking for work.
   b. to persons of legal age without jobs who are seeking work.
   c. only to persons who are not looking for employment.
   d. to children and older persons who are not able to work.
   e. primarily to temporary and part-time workers.

12. A primary outcome of long-term, chronic unemployment is
   a. good health.
   b. an increased chance of being able to obtain part-time work.
   c. upward social mobility.
   d. poverty.
   e. an increase in the odds that a person will further their education.

13. U.S. workers report high degrees of job satisfaction. Recent *General Social Survey* data indicate that only __________ of workers indicate that they are very dissatisfied with their jobs.
   a. 1 %
   b. 3 %
   c. 5 %
   d. 10 %
   e. 15 %

14. The seeds of the mass education movement in the U.S. were planted by
   a. Spanish explorers.
   b. French traders.
   c. Thomas Jefferson.
   d. the Puritans.
   f. Christopher Columbus.

15. Cross-national data reveal that the average number of years of schooling is greatest in
   a. France.
   b. Japan.
   c. Cuba.
   d. China.
   e. the United States.

16. Education refers to
   a. what a person has learned.
   b. the time spent in school.
   c. a and b
   d. public and private schools.
   e. homeschooling.

17. The 2002 *Knowledge and Skills for Life* study reveals that academic achievement among fifteen-year-olds is highest in
   a. the United States.
   b. Japan.
   c. Great Britain.
   d. Brazil.
   e. Italy.

18. Which of the following statements about the Educational Testing Service study on U.S. literacy is false.
   a. The dimensions of literacy tested were quantitative, documentary and prose.
   b. Approximately half of the respondents scored low on each literacy dimension.
   c. Only 4% of the respondents were able to perform at the highest level of quantitative literacy.
   d. The majority of respondents with college degrees scored in the lower ranges on quantitative and prose literacy.
   e. Each statement is correct.

19. Coleman observed that student achievement is more likely to be influenced by
   a. the size of the school library.
   b. teacher competence.
   c. class size.
   d. differences in per pupil educational expenditures.
   e. age of the teaching facility.

20. Heyns' study suggests that schools may be more effective if
   a. the school year is lengthened.
   b. the school year is shortened.
   c. the school day is lengthened.
   d. the school day is shortened.
   e. more emphasis is placed on science and math.

21. The Alexander, Natriello and Pallas study shows that cognitive development is higher among
   a. students who stay in school.
   b. high school dropouts.
   c. high school students enrolled in public schools.
   d. high school students enrolled in private schools.
   e. high school students who choose to homeschool.

22. Data from the "High School and Beyond" study indicate that the percentage of high school dropouts is highest among
   a. Native Americans.
   b. Hispanic Americans.
   c. Whites.
   d. Asian Americans.
   e. African Americans.

23. Compared to students enrolled in Catholic and private schools, students enrolled in pubic schools are
   a. less likely to drop out.
   b. more likely to study.
   c. less likely to report experiencing discipline problems.
   d. more likely to plan to go to college.
   e. none of the above

24. Recent studies indicate that children who homeschool
   a. are primarily from affluent families.
   b. are able to do so because both parents work.
   c. receive an inferior education.
   d. are not able to take field trips.
   e. perform better on standardized tests.

25. The cross-national study on schooling by Heyneman and Loxley indicates that the positive correlation between income and education is
    a. higher among poorer nations.
    b. the same for poor and rich nations.
    c. higher among richer nations.
    d. is stronger for female workers.
    e. is stronger for younger workers.

26. The decline in the value of a college education has been attributed to
    a. an increase in the cost of college education.
    b. a decrease in the cost of college education.
    c. an increase in the supply of college graduates.
    d. a decrease in the supply of college graduates.
    e. the low literacy levels of college graduates.

27. Mean annual income levels for persons aged 45-55 are higher for persons who
    a. have a master's degree or more.
    b. are college graduates.
    c. have some college or technical training beyond high school.
    d. are high school graduates.
    e. never completed high school.

28. In the "credential society" educational credentials
    a. control the supply of workers in many occupations.
    b. increase a worker's replaceability.
    c. decrease a worker's replaceability.
    d. a and b
    e. a and c

29. These theories argue that the primary function of education is to place students in the stratification system.
    a. literacy theories
    b. schooling theories
    c. allocation theories
    d. scientific management theories
    e. occupational prestige theories

30. Meyer maintains that
    a. educational levels reflect distinct social statuses.
    b. college graduates share more in common with each other than with persons who are not college graduates.
    c. educational attainment statuses continue to impact a person's identity and quality of life throughout life.
    d. educational systems have the power to legitimate new knowledge areas and new social statuses.
    e. all of the above

True/False

Answers and page references are provided at the chapter end.

1. Most of the time higher education and higher income are associated with higher occupational status.

2. Lawyer is to nightclub singer as low occupational prestige is to high occupational prestige.

3. Technological innovations have increased worker productivity for the most part by enabling persons to work smarter rather than harder.

4. U.S. labor force participation rates are high. Recent figures show that two-thirds of the population aged 16 and over is involved in the labor force.

5. Poverty is a major outcome associated with chronic unemployment.

6. Schooling measures academic performance while education addresses the amount of time spent in school.

7. According to the ETS study on literacy, two-thirds of the U.S. residents surveyed scored high on the three dimensions of literacy.

8. Coleman's research on the quality of U.S. education suggested that the quality of educational facilities does not exert a substantial impact on student achievement.

9. According to the "High School and Beyond" data, whites, compared to other racial and ethnic groups, are more likely to study and less likely to drop out of school.

10. Homeschooling is not as popular today as it was ten years ago.

Short Answer Questions

These short answer questions are provided to test your knowledge and understanding of the basic sociological concepts presented in Chapter 16. Page references for answers are included at the chapter end.

1. Indicate how Hatt and North computed occupational prestige and provide examples of high and low prestige occupations.

2. Identify five factors associated with the increase in female labor force participation.

3. What criteria are used in defining unemployment?

4. Identify the three dimensions of literacy measured in the Educational Testing Service study. How did most U.S. residents score on these dimensions?

5. According to Heyns' research, what factor had the greatest impact on children's learning during the summer school recess months?

6. What is the "High School and Beyond Study," and why is it an important source of educational data?

7. Identify three objections to homeschooling offered by the educational establishment.

8. How has the college degree been devalued in contemporary U.S. society?

9. Collins claims that the expansion of higher education in the U.S. has created a "credential society." What does this mean?

10. What is the nature of the association among educational attainment, status identification and the creation of new social statuses?

Essay Questions

These questions are designed to test your understanding of key sociological concepts presented in Chapter 16 and your ability to apply these insights to concrete situations.

1. Some sociologists argue that employment in the manufacturing sector of the economy is declining while employment in the information and service sectors is increasing. How might these shifts in the structure of the work place alter occupational prestige scores?

2. Describe how "scientific management" principles have been applied to the delivery of mass education in contemporary U.S. society.

3. Year-round schools are becoming increasingly more popular. Debate whether or not year-round schools will significantly impact student achievement levels.

4. An increasing number of children are being homeschooled. What are some of the advantages and disadvantages of homeschooling, and how might homeschooling impact public and private schools?

5. Collins argues the U.S. is a "credential society." If this is the case, is U.S. society characterized by an educational caste system?

Answers

Multiple Choice

1. d (449)
2. a (450)
3. b (451)
4 e (452)
5. c (454)
6. d (454)
7. b (455-456)
8. e (455)
9. c (456)
10. a (456)
11. b (456)
12. d (456)
13. b (457)
14. d (458)
15. e (458)
16. a (459)
17. b (459-460)
18. e (460-461)
19. b (461-462)
20. a (463)
21. a (463)
22. b (464)
23. e (465)
24. e (466)
25 a (467)
26. c (467)
27. a (468)
28. e (468-469)
29. c (469)
30. e (469-471)

True/False

1. T (449)
2. F (451)
3. T (452)
4. T (454)
5. T (456)
6. F (459)
7. F (460-461)
8. T (461)
9. F (464)
10. F (466)

Short Answer

1. (450-451)
2. (455-456)
3. (456)
4. (460-461)
5. (462-463)
6. (463-465)
7. (466)
8. (467-468)
9. (468-469)
10. (469-471)

CHAPTER 17

SOCIAL CHANGE: DEVELOPMENT AND GLOBAL INEQUALITY

Extended Abstract

Forces promoting stability and change influence social structures. Change may be derived from internal and external sources, and different aspects of social life change at different rates. This chapter addresses sources of social change, presents the modernization process as an example of social and economic change and assesses the differing impact of technological change on global development.

Three primary internal sources of social change are introduced. These are innovations, conflict and population growth. Innovations include new technology like the steam engine, new beliefs and practices such as the idea of progress, or new social structures like changing gender roles. Social movements within a society often arise when groups experience inequality, and may generate conflict, which may bring about social change. The Civil Rights Movement is cited as an example. The third internal source of social change is growth. Population growth is associated with many forms of structural change, such as representative government and urbanization.

Changes may be introduced to a society at one point in time, but society's response may be more immediate or delayed. A delayed response to change is known as cultural lag. During periods of cultural lag, societies may experience higher degrees of internal conflict. For example, machines may reduce the need for a large work force. This does not mean that persons will voluntarily limit fertility in the near future. Since fertility patterns are often legitimated by deeply held cultural traditions, fertility regulation could generate conflict.

External sources of social change include diffusion, conflict and environmental constraints or ecological factors. Diffusion addresses the borrowing and sharing of ideas and technology. Social scientists generally agree that the majority of a society's cultural content has been obtained through diffusion. The diffusion process may be slow or rapid. The diffusion of TV in the U.S. and throughout the world is provided as an example. Threats of attack from other societies are examples of conflict as an external source of change. Finally, ecological factors such as natural disasters and climate change are external sources of social change. The Vikings' failure to adapt to climate changes in Artic regions is an example of how environmental constraints impact social change.

Four theories addressing the rise of capitalism and economic development are developed next. These theories are the Marxian critique of capitalism, Weber's Protestant Ethic, the taming of the state theory and dependency theory. Marx argued that modernization and industrial development are products of capitalism. Capitalism stimulates productivity due to the focus on self-interest and the free market. Under this system persons are rewarded for being more productive. However, with the private control of the means of production, individuals exploit others rather than share the wealth created. A communist revolution would be required to insure that the wealth created under capitalism would be shared. Contrasting capitalist and pre-capitalist societies, Marx noted that pre-capitalist, command economies were ineffective because they did not motivate people to produce more. Under a command economy, leaders control production and receive the benefits of the labor of others.

Weber's "Protestant Ethic" thesis offers an explanation for the rise of capitalism. Based on his analysis of the writings of Calvin, a Protestant reformer, Weber maintained that Protestantism encouraged valuing work, limiting consumption and reinvesting available resources. These value orientations would enable persons to accumulate wealth, which would stimulate the development of capitalism and the Industrial Revolution.

According to the state theory of modernization, the state must be controlled because ruling elites (command economies) exploit the masses through overtaxation and the seizure of property. Since people will not develop new technology if they cannot benefit from it, capitalism and industrial development are contingent upon the creation of a free market and the "taming" of the state.

The fourth theory is world system-dependency theory. According to this theory, rich nations maintain their power and standard of living by exploiting poor nations. Poor nations are restricted in developing their industrial potential and send their raw materials to rich nations. The outcome is a stratification of nations (core-semiperipheral-pheripheral) within the world system. To test the claims of dependency theory, Delacroix analyzed changes in per capita gross domestic product (GDP) and changes in educational attainment among fifty-nine nations over the 1955-1970-time period. Delacroix observed that neither variable was influenced by the export of raw materials measure. However, a strong positive association was noted between educational attainment and per capita gross domestic product. This suggests that modernization may be influenced more by internal rather than external (world system) forces. The association between educational attainment and economic development is confirmed by a more recent cross-national study by Firebaugh and Beck.

The magnitude of global inequality is outlined briefly in the last section. First, substantial differences in per capita gross national income (GNI) exist between more developed and less developed nations. Second, life expectancy varies by as much as 25 to 35 years as poorer nations lack adequate medical facilities and consume poor diets. Proper sanitation is also a major problem among poorer nations, and wealth tends to be more concentrated within a local rather than a foreign elite group. In concluding this chapter, Stark discusses the globalization process and the "global village." While there is evidence that global communication and economic networks are being formed, a unified global culture seems improbable. Individual cultures tend to be grounded in local networks that promote solidarity. Therefore, "global villages" may not be an inevitable outcome of global communication and economic exchange.

Key Learning Objectives

After a thorough reading of Chapter 17, you should be able to:

1. distinguish internal and external sources of social change.

2. provide examples of innovations.

3. note how conflict can be an internal and external source of social change.

4. describe how population change can function as an internal source of social change.

5. define cultural lag and provide examples.

6. explain why the majority of a society's cultural content is derived from diffusion.

7. demonstrate how environmental (ecological) factors can bring about social change.

8. identify key aspects of the modernization process.

9. outline the major tenets of capitalism.

10. articulate the Marxian critique of capitalism.

11. state how workers are exploited under a command economy.

12. indicate how the Protestant Reformation could stimulate the growth of capitalism.

13. critically evaluate world system-dependency theory.

14. identify the distinguishing characteristics of core, semi-peripheral and peripheral nations.

15. describe the key elements associated with the globalization process and comment on the likelihood of attaining a unified "global village."

Chapter Outline

I. Internal Sources of Social Change and Cultural Lag
   A. Innovations: The Power of New Ideas
      1. New ideas stimulate technological developments, which bring about change.
      2. Beliefs, like the "idea of progress," can promote social change.
      3. Innovations can lead to the creation of new social structures, such as redefined gender roles.
   B. Conflict and Social Change: The Impact of Social Movements
   C. Population Growth and Structural Change: Urbanization and Representative Government
   D. Cultural Lag: Gauging the Pace of Change within a Society
      1. A society's response to change can be slow.
      2. Slow responses to change can generate conflict.

II. External Sources of Social Change
   A. Diffusion: The Borrowing and Sharing of Ideas and Technology
   B. External Threats and Conflict
   C. Ecological Sources
      1. Natural resource depletion, natural disasters and climate variations

can stimulate social change.

2. The Vikings were unable to adapt to climate change in Greenland.

III. Modernization and Capitalism

A. Modernization: The Agrarian-Industrial Transformation

B. The Marxian Critique of Capitalism

1. Capitalism generates modernization and industrial development.
2. Within a capitalist economy, self-interest and the free market stimulate productivity.
3. Private ownership of the means of production leads to concentration of wealth.
4. Communist revolutions insure that the wealth created through capitalism is shared.
5. Precapitalist command economies failed because they did not stimulate worker productivity.

C. The Protestant Ethic and Capitalism

1. Protestantism placed a high value on work and stressed the limited consumption of resources and the reinvestment of savings.
2. Protestant value orientations provided an environment conducive to the spreading of capitalism.

D. The State Theory of Modernization

1. Command economies (ruling elite) exploit the masses through overtaxation and property seizure.
2. Command economies suppress the development of new technology.
3. Capitalism is contingent upon the taming of the state and the development of a free market.

E. World System-Dependency Theory

1. Rich nations exploit poor nations.
2. Poor suffer from restrictions placed on industrial development and send raw material to rich nations.
3. A global system of stratification exists (core nations, semiperipheral nations, peripheral nations).
4. Delacroix conducted a cross-national study to test the claims of dependency theory.
    a. Discovered that export of raw materials does not impact change in gross domestic product or change in educational attainment.
    b. A strong link exists between change in educational attainment and change in gross domestic product.
    c. Findings suggest that modernization may be influenced more by internal rather than external factors.
5. Firebaugh and Beck's study confirms strong impact of educational attainment on economic development.
6. The impact of education on industrialization is strong if the quality of the education is standardized, a large audience is reached and the training is relevant.

7. Inequality tends to be domestically based and may be the result of resource mismanagement, poor leadership and political and economic corruption.
8. Global inequality appears to be declining.

IV. Global Inequality and Globalization
   A. Documenting Global Inequality
      1. The variations in per capita gross national income between rich and poor nations is extensive.
      2. The life expectancy gap between rich and poor nations can be as large as 25 to 35 years.
      3. Residents of poorer nations are more likely to encounter inadequate health care, poor diets and poor sanitation.
   B. Globalization and the "Global Village"
      1. The globalization process involves the building of unified communication and economic networks and the creation of a unified culture.
      2. Progress has been made toward the creation of global communication and economic networks, but the creation of a unified culture is less likely.
      3. Individual cultures are grounded in local networks that promote solidarity.

Key Terms

Based on your reading of Chapter 17, you should be able to define and illustrate the key sociological concepts listed below. Page numbers are provided in parentheses as reference points.

Modernization (478)
Innovations (479)
Idea of Progress (480)
Cultural Lag (482)
Social Evolution (483)
Diffusion (483)
Capitalism (488)
Free Market (488)
Command Economies (488)
Protestant Reformation (489)
Spirit of Capitalism (490)
Protestant Ethic (490)
State Theory of Modernization (491)
World System Theory (492)
Dependency Theory (492)
Core Nations (493)
Peripheral Nations (494)
Semiperipheral Nations (494)
Gross Domestic Product (495)
Gross National Income (498)
Globalization (500)
Global Village (500)

Key Research Studies

Listed below are key research studies cited in Chapter 17. Familiarize yourself with the major findings of these studies. Page references are provided in parentheses.

Delacroix: tested the claims of dependency theory with cross-national data and observed that export of raw materials does not influence change in per capita gross domestic product or change in educational attainment. A link is discovered between change in educational attainment and change in per capita gross domestic product (495-497).

Hage, Garnier and Fuller: discovered that education impacts industrialization if the quality of the education offered is standardized, a large number of people are provided an opportunity to receive an education and the educational system provides relevant training (497).

Firebaugh and Beck: tested claims of dependency theory with cross-national data and found little support for dependency theory. They observed that national quality of life is influenced more by internal economic development and that secondary education stimulates economic development (498).

Bradshaw and Wallace: inequality tends to be domestically based and is frequently a result of the mismanagement of resources, poor leadership and political and economic corruption (499).

Firebaugh: study indicates that global income inequality has been declining in recent years (500).

McLuhan: developed concept of the "global village." Globalization involves the building of unified communication and economic networks and the creation of a unified culture. (510-511).

Info Trac Search Words

Enter these search terms to conduct more extensive investigations of key topics introduced in Chapter 17.

Culture Diffusion

Economic Development
Subdivisions: Case Studies
Select one of the case studies listed. Note how economic development is measured and tested.

Protestant Ethic
Review the studies cited and select one. What aspect of the Protestant Ethic thesis is being tested? Do the study findings support or challenge the Protestant Ethic thesis?

Gross Domestic Product
Subdivisions: Economic Aspects

Globalization
Subdivisions: Social Aspects

Multiple Choice

Answers and page references are provided at the chapter end.

1. The raincoat, dry cleaning, the ice-cream cone and the ballpoint pen are all examples of
   a. communism.
   b. feudalism.
   c. modernization.
   d. ecological change.
   e. the Protestant ethic.

2. Stark argues that new technology, new culture and new social structures are examples of
   a. innovation.
   b. diffusion.
   c. population change.
   d. external conflict.
   e. ecological change.

3. Conflict can
   a. be an internal source of social change.
   b. be an external source of social change.
   c. be more frequent during periods of cultural lag.
   d. all of the above
   e. none of the above

4. The *World Values Survey* data indicate that persons residing in __________ are least likely to view scientific advances as being helpful in the long run.
   a. the United States
   b. Japan
   c. Germany
   d. Turkey
   e. Nigeria

5. Broadband Internet Access has tended to spread from more densely populated areas to less densely populated areas. This is an example of
   a. an internal source of conflict.
   b. population change as a source of social change.
   c. innovation.
   d. ecological change as a source of social change.
   e. diffusion.

6. The borrowing and sharing of innovations is known as
   a. cultural lag.
   b. diffusion.
   c. modernization.
   d. ecological change.
   e. globalization.

7. TV sets per 1,000 population is highest in
   a. the United States.
   b. France.
   c. Japan.
   d. Sweden.
   e. Spain.

8. Droughts, natural disasters and the depletion of natural resources would be examples of this source of social change.
   a. diffusion
   b. conflict
   c. population growth
   d. ecological
   e. innovation

9. According to Marx
   a. modernization was the result of capitalism.
   b. capitalism was the result of modernization.
   c. the Protestant Reformation led to the development of capitalism.
   d. the development of capitalism led to the Protestant Reformation.
   e. command economies replaced communism.

10. Marx believed that workers were motivated by
   a. religious beliefs.
   b. political ideologies.
   c. public opinion.
   d. self-interest.
   e. concern for the common good.

11. Capitalism is an economic system based on
   a. the ruling elite.
   b. the public ownership of the means of production.
   c. the private ownership of the means of production.
   d. the principle of taxation without representation.
   e. the preservation of the coercive state.

12. In a free market, prices and wages are determined by
   a. business leaders.
   b. consumers.
   c. the state.
   d. religious authorities.
   e. bankers.

13. The power of an emperor or lord would be maximized in a
   a. free market.
   b. tamed state.
   c. command economy.
   d. all of the above
   e. none of the above

14. In a command economy, consumption
   a. and surplus production are high.
   b. and surplus production are low.
   c. is low and surplus production is high.
   d. is high and surplus production is low.
   e. equals surplus production.

15. Marx believed that capitalism was
   a. invented by the bourgeoisie.
   b. invented by the proletariat.
   c. the outcome of the Protestant Reformation.
   d. the outcome of communism.
   e. created by the military.

16. The Protestant Ethic thesis was developed by
   a. Marx.
   b. Weber.
   c. Durkheim.
   d. Delacroix.
   e. Wallerstein.

17. According to the Protestant Ethic thesis, Protestantism motivated people to
   a. reinvest savings to increase wealth.
   b. increase personal consumption.
   c. decrease personal consumption.
   d. a and b
   e. a and c

18. Calvin believed that __________ was a sign of salvation.
   a. a successful life
   b. poverty
   c. becoming a religious leader
   d. becoming a military officer
   e. becoming a political leader

19. The belief that economic success reflects God's grace is a key idea associated with
   a. command economies.
   b. dependency theory.
   c. the idea of progress.
   d. the Protestant Ethic.
   e. communism.

20. According to the state theory of modernization, capitalism flourishes
   a. under a coercive state.
   b. when coercive states are "tamed."
   c. when the state regulates religion.
   d. when the state deregulates religion.
   e. once a communist state is created.

21. Which of the following statements is false?
   a. In a command economy, the state supports itself by confiscating surplus production.
   b. In a command economy, surplus production is consumed primarily by the ruling elite.
   c. In a command economy, people are free to pursue their economic self-interests.
   d. In a command economy, people do not develop new technology, because they cannot benefit from it.
   e. each statement is false

22. World system-dependency theory is often associated with
   a. Wallerstein.
   b. Weber.
   c. Durkheim.
   d. Calvin.
   e. Linton.

23. Dependency theory maintains that
   a. stratification exists among nations.
   b. rich nations exploit poor nations.
   c. economic and political relations among nations unite them into a single social system.
   d. rich nations dominate poor nations by controlling trade relationships.
   e. all of the above

24. The most powerful nations in the world system are part of the
   a. bourgeoisie.
   b. proletariat.
   c. core.
   d. semiperiphery.
   e. periphery.

25. In the world system, nations in the __________ have highly specialized economies, a higher degree of political instability and class conflict, and supply raw materials to more powerful nations
   a. periphery
   b. semiperiphery
   c. core
   d. proletariat
   e. bourgeoisie

26. In testing the claims of dependency theory, Delacroix discovered that export of raw materials
   a. exerts a strong impact on gross domestic product only.
   b. exerts a strong impact on educational attainment only.
   c. cannot be measured adequately.
   d. does not impact gross domestic product or educational attainment.
   e. impacts both gross domestic product and educational attainment.

27. Gross domestic product (GDP) is
   a. a measure of labor supply.
   b. a measure of labor demand.
   c. the market value of all goods and services produced in a given year.
   d. the total value of government services.
   e. the total value of all import goods in a given year.

28. The cross-national study by Firebaugh and Beck indicates that
   a. improvements in secondary education strengthen economic development.
   b. improvements in secondary education restrict economic development.
   c. secondary education does not impact economic development.
   d. the export of raw materials impacts per capita gross domestic product.
   e. the export of raw materials impacts educational attainment.

29. According to 2002 *World Bank* data, __________ is currently the nation with the highest per capita gross domestic product.
   a. Mexico
   b. Switzerland
   c. Ethiopia
   d. Japan
   e. the United States

30. Presently life expectancy is highest in
   a. Zimbabwe.
   b. Japan.
   c. Poland.
   d. the United States.
   e. India.

True/False

Answers and page references are provided at the chapter end.

1. The modernization process involves the transformation of agrarian societies into industrial societies.

2. The "idea of progress" is an example of an innovation.

3. According to Stark conflicts can be examples of internal or external sources of social change.

4. Mechanized agriculture may reduce the need for large families. This does not mean that families will automatically start to reduce household size. This is an example of command economies.

5. Marx argued that the key elements of capitalism are self-interest, the private ownership of the means of production and the free market.

6. Weber argued that capitalism provided the foundation for the spread of the Protestant Reformation.

7. According to the state theory of modernization, patent laws reduce the coercive power of the state and stimulate technological development.

8. According to dependency theory, rich, industrial nations are classified as periphery nations.

9. Delacroix's research indicates that the export of raw materials exerts a significant influence on a nation's gross domestic product.

10. Per capita gross national income and life expectancy are higher in nations normally classified as peripheral nations.

Short Answer Questions

These short answer questions are provided to test your knowledge and understanding of the basic sociological concepts presented in Chapter 17. Page references for answers are included at the chapter end.

1. Indicate why modernization is an appropriate example of social change.

2. Provide examples of three internal sources of social change.

3. Why are periods of cultural lag associated with a higher risk of internal conflict?

4. Why is diffusion an important external source of social change?

5. Identify the key elements of the Marxian critique of capitalism.

6. Why did Weber believe that the Protestant Reformation provided the roots for the development and expansion of capitalism?

7. Explain why a free market is not likely to develop in a command economy.

8. Distinguish core, semiperiphery and periphery nations.

9. What is gross domestic product (GDP), and why is it cited as an indicator of global inequality?

10. Identify three major characteristics of globalization.

Essay Questions

These questions are designed to test your understanding of key sociological concepts presented in Chapter 17 and your ability to apply these insights to concrete situations.

1. Explain how population growth could be an example of an internal and an external source of social change.

2. How could communist regimes be conceptualized as contemporary examples of command economies?

3. Describe how the "Protestant Ethic" hypothesis could be tested using cross-national data.

4. Recent studies indicate that the correlation between educational attainment and economic development is strong. Why is this the case?

5. Would the creation of a "global village" increase or reduce global inequality? Justify your answer.

Answers

Multiple Choice

1. c (479)
2. a (479-480)
3. d (480-481, 484-485)
4. b (481)
5. e (483)
6. b (483)
7. a (484)
8. d (485-486)
9. a (487)
10. d (488)
11. c (488)
12. b (488)
13. c (488-489)
14. d (489)
15. a (489)
16. b (489)
17. e (489)
18. a (490)
19. d (490, 502)
20. b (491)
21. c (491-492)
22. a (492)
23. e (492-494)
24. c (493-494)
25. a (494)
26. d (495-497)
27. c (495)
28. a (498)
29. b (498)
30. b (499)

True/False

1. T (478)
2. T (479-480490)
3. T (480-481; 484-485)
4. F (481-482)
5. T (487-489)
6. F (489-491)
7. T (492)
8. F (493-494)
9. F (495-497)
10. F (498-499)

Short Answer

1. (477-479)
2. (479-481)
3. (481-482)
4. (483-484)
5. (487-489)
6. (489-491)
7. (488-489, 491-492)
8. (493-494)
9. (495-498)
10. (500)

CHAPTER 18

POPULATION CHANGES

Extended Abstract

Demography involves the study of human population characteristics and population change. The major components of population change are fertility, mortality and migration. Censuses, which are total population counts, have been conducted throughout history. Egyptian pharaohs and Roman emperors conducted censuses for taxation purposes and to estimate military potential. William the Conqueror's *Domesday Book* (1066) is an example of a massive population count. Population censuses, however, are expensive and must be updated.

Demographers utilize various methodological techniques to gauge and portray population characteristics. Three techniques commonly employed are the comparison of rates, cohort analysis and age and sex structures (population structures). Rates can involve ratio or percentage measures, like the annual population growth rate, or they may involve measures based on standardized population units, such as per 1,000 population. The crude birth rate and the crude death rate are examples of the latter. Demographers must be careful when crude measures are used because the rates can be distorted by the population's age structure. For example, the crude birth rate is the number of births per 1,000 population. The measure could vary substantially among different populations because the percentage of women of reproductive age (aged 15-44) is not the same. In studying fertility demographers counter this problem by utilizing more direct measures of fertility such as the fertility rate, the average number of births to women of reproductive age. Age-specific fertility and mortality rates are also calculated to specify more subtle population patterns.

Cohort analysis enables researchers to track demographic changes among a group of persons born during a particular time period. A well-known cohort is the baby boom generation. From 1946 to the mid-1960s, fertility in the U.S. increased significantly. Children born during this period are known as baby boomers. Stark provides a special insert indicating how the baby boom generation has impacted suburban growth, educational needs and consumer tastes. The anticipated impact of the baby boom generation on retirement and possible labor shortages is noted also.

Population researchers use age and sex structures to identify patterns of population growth, stability and decline. Age and sex structures document population distribution by age group and gender composition within each age group. Demographers distinguish three basic age and sex (population) structures. These are expansive, stationary and constrictive population structures. Expansive population structures portray younger, growing populations. Younger cohorts are larger than older cohorts. Stationary population structures reflect a stable population. A population is neither growing nor declining. Younger cohorts equal older cohorts. Constrictive population structures describe older, declining populations. Younger cohorts are smaller than older cohorts.

The remainder of the chapter is devoted to a discussion of six major shifts in population trends throughout human history. Stark employs this framework to introduce important demographic theories developed by Malthus and by Davis (the demographic transition). The first population shift occurred approximately 10,000 years ago and is associated with the agricultural revolution and the development of agrarian societies. Prior to the Industrial Revolution, fertility rates in non-industrial societies were high, but mortality levels were high also due to famine, disease and war. This resulted in low population growth levels. Thomas Malthus' theory of population change stressed the impact of mortality on population control. Developing an insight noted by Adam Smith, Malthus linked population growth to the available food supply. Smith had argued that populations expand in proportion to the availability of necessary life-supporting resources. Malthus, in his *Essay on the Principle of Population* (1798), argued that population tends to grow at an increasing, exponential rate while the growth of the food supply is more limited. In the short run, populations may experience growth, but increasing population density and food shortages will inevitably limit population growth as mortality levels begin to rise. Thus, famine, disease and war function are positive checks that restrict continued, rapid population growth. Malthus also argued that fertility levels would remain high and would not check population growth.

The second great population shift occurred in the 1800s with the European Industrial Revolution. The mechanization of agriculture provided an environment that stimulated the Industrial Revolution. The mechanization of agriculture meant that fewer persons would be needed to work the land. This simulated rural-urban migration. The mechanization of agriculture also enabled farmers to increase food production. Populations began to grow, as more food was available. Modernization and the Industrial Revolution also brought about other changes that would impact population growth. These changes included factors that would lower mortality like better health care, improved diet and better sanitation. Consequently, increases in the food supply and declining mortality stimulated population growth. Malthus had argued that population control would be possible only if mortality levels were to rise. This did not happen.

The industrial nations of Europe did eventually begin to experience a decline in population growth, but it came from a decline in fertility. This fertility decline signaled the third major shift in population trends. According to Davis, the decline in fertility was associated with the increased cost associated with maintaining large families in industrial societies. As a rational response to this increased cost, couples eventually choose to limit fertility. This leads to a realignment of fertility and mortality at lower levels and to a reduction in the population growth rate. Davis' demographic transition theory describes a society's transition from a high fertility – high mortality environment to a low fertility – low mortality environment. The initial explosion in population associated with the mechanization of agriculture and the beginning stages of industrialization is attributed to the fact that mortality rates fall before fertility rates. As the perceived cost of children increases, couples choose to reduce fertility, and this results in the realignment of fertility and mortality at lower levels. This realignment produces lower rates of population growth. Researchers have demonstrated that fertility declines are more likely to occur once certain modernization thresholds are met. These thresholds address important factors such as declining agricultural labor force participation, increased educational attainment and literacy, increased life expectancy, declines in infant mortality, delay in the age of first marriage and improved standard of living.

The fourth major shift in demographic trends was somewhat surprising and has had a substantial impact on global population growth. This shift occurred during the mid-twentieth century and has involved the explosion of population in developing countries. As developed countries have introduced mechanized agriculture, public health and industrialization in developing countries, mortality rates have declined sharply. Consequently, developing nations have had little time to respond to the resulting rapid increase in population growth.

The fifth major population shift is associated with the evidence of a fertility reversal in some developing nations. Continued economic development is contingent upon population control. As the perceived cost of children increases, fertility tends to fall. A brief analysis of available Chinese data indicates that declining fertility is associated with lower agricultural employment levels, higher levels of educational attainment, improved life expectancy and increasing affluence. It is important to note that China's fertility decline began before coercive fertility control measures were introduced. Recent studies of wanted fertility in developing countries suggest that women are giving birth to more children than they desire. Some researchers have linked the high level of unwanted births to difficulties encountered in obtaining contraceptives.

The sixth major shift in population trends is associated with the decline in fertility in many industrial countries to below replacement levels. Replacement-level fertility is between 2.02 to 2.07 births. The rate is slightly above 2 due to variations in infant mortality. The fertility rate for many industrialized nations is below 2.0 births. This decline in fertility has raised concern among some population researchers. In particular, what impact will continued declining population growth have upon future labor needs? Will these countries experience an increase in the immigration of people of working age? How will retirement and quality of life be affected? Finally, will low fertility levels stimulate gender-based birth preferences?

Key Learning Objectives

After a thorough reading of Chapter 18, you should be able to:

1. describe early Egyptian, Roman and English attempts to provide comprehensive population counts.

2. identify the three major components of population change.

3. explain how measures based on crude rates can be misleading.

4. state why demographers employ cohort analysis.

5. specify which population groups are included in the baby boom cohort, and indicate how the baby boom generation has and will influence U.S. life.

6. distinguish the three basic types of population structures.

7. describe key events associated with each of the six shifts in population trends.

8. identify three factors contributing to high mortality in preindustrial societies.

9. explain Malthus' proposed association between population growth and the food supply.

10. state the key tenets of the demographic transition theory.

11. identify important modernization thresholds associated with fertility reduction.

12. explain why the demographic transition in developing countries has been more rapid.
13. identify important factors associated with the fertility decline in developing countries.

14. distinguish measures of numeracy and wanted fertility.

15. indicate problems that could be encountered by nations experiencing below replacement level population growth.

Chapter Outline

I. Demography and Basic Demographic Tools
   A. Demography: The Study of Human Population
      1. Societies conduct censuses to obtain a total population count and gain information that can be used for future planning.
      2. The three major population dynamics are fertility, mortality and migration.
   B. Crude Rates: Standardized Population Units
      1. Crude measures of fertility and mortality are based on standard population units such as per 1,000 population.
      2. Crude rates can be influenced by a population's age structure.
      3. Age-specific measures and other more direct measures of fertility and mortality are used to specify more subtle population patterns.
   C. Cohort Analysis: Tracking Demographic Change
      1. A cohort includes all persons born at a particular time.
      2. Cohorts represent distinct demographic groups that can be followed throughout the life cycle.
      3. The baby boom generation is large U.S. birth cohort involving persons born between 1946 and the mid-1960s.
      4. The baby boom generation has exerted an impact on suburban growth, educational needs and consumer tastes and will impact future retirement and labor needs.
   D. Age-Sex (Population) Structures and Population Growth
      1. Expansive population structures are associated with high population growth rates and large, younger age cohorts.
      2. Stationary population structures reflect stable populations. The population is neither growing nor declining. Younger and older cohorts are of more equal size.
      3. Constrictive population structures are associated with declining population growth and are characterized by larger, older cohorts.

II. Six Major Shifts in Population Trends
   A. The First Shift: Declining Mortality and Preindustrial Societies
      1. Prior to the industrial revolution, population growth was low because fertility and mortality rates were high.

2. The major sources of high mortality were famine, disease and war.
3. Malthus developed a population model linking population growth and the food supply.
    a. Population tends to grow exponentially while the food supply growth is more restrictive.
    b. Short-term population increases are controlled by increases in mortality (positive checks).
    c. Fertility rates remain high and are unlikely to be lowered to check population growth.

B. The Second Shift: The European Industrial Revolution
1. The Industrial Revolution was stimulated by the mechanization of agriculture.
2. The mechanization of agriculture enabled farmers to produce more food and reduced the need to maintain large families.
3. Increasing modernization was also associated with better health care, improved diet and better sanitation. These factors lower mortality.
4. The increase in population growth experienced in the early stages of the Industrial Revolution was the result of declining mortality.

C. The Third Shift: The European Demographic Transition
1. Demographic transition theory describes a society's transition from a high fertility - high mortality environment to a low fertility - low mortality environment.
    a. Rapid population growth occurs when mortality rates fall before fertility rates.
    b. The population growth rate begins to decline as the fertility rate declines.
    c. The decline in fertility is linked to the increasing cost of children.
2. Researchers have demonstrated that fertility decline is more likely when certain modernization thresholds are met
3. Important fertility reduction thresholds include declining agricultural labor force participation, increased education and literacy, increasing life expectancy, declining infant mortality, a delay in the age of first marriage and improved standard of living.

D. The Fourth Shift: Rapid Population Growth in Developing Nations
1. Rapid population growth among developing nations began in the mid-twentieth century.
2. A rapid decline in mortality has been experienced as technological innovations from developed countries have been adopted.
3. Developing nations have had little time to respond to the high rates of population growth generated by the rapid mortality decline.

E. The Fifth Shift: Fertility Decline in Developing Nations
1. Continued economic development is contingent upon population control.
2. As the perceived cost of children increases, fertility declines.
3. Declining fertility in China is correlated with lower agricultural

employment levels, higher educational attainment, improved life expectancy and increasing affluence.

4. Studies on wanted fertility indicate that women in developing countries are giving birth to large numbers of unwanted children.

F. The Sixth Shift: Below Replacement-Level Fertility and Industrial Nations

1. Replacement-level fertility is between 2.02 and 2.07 births.
2. Current fertility levels among many industrial nations is under 2.0 births.
3. Below replacement-level fertility generates concern over such issues as labor shortages, the need for increased immigration, retirement benefits and gender-based birth preferences.

Key Terms

Based on your reading of Chapter 18, you should be able to define and illustrate the key sociological concepts listed below. Page numbers are provided in parentheses as reference points.

Domesday Book (505)
Census (506)
Demography (507)
Growth Rate (507)
Crude Death Rate (508)
Crude Birth Rate (508)
Fertility Rate (508)
Age-Specific Death Rates (508)
Birth Cohort (509)
Age Structure (510)
Sex Structure (510)
Expansive Population Structure (510)
Stationary Population Structure (512)
Constrictive Population Structure (512)
Arithmetic Increase (515)
Exponential Increase (515)
Positive Checks (515)
Malthusian Theory of Population (516)
Replacement-Level Fertility (518)
Zero Population Growth (518)
Demographic Transition (518)
Demographic Transition Theory (520)
Thresholds of Modernization (520)
Baby Boom (524)
Numeracy (528)
Wanted Fertility (529)
Depopulation (530)

Key Research Studies

Listed below are key research studies cited in Chapter 18. Familiarize yourself with the major findings of these studies. Page references are provided in parentheses.

Domesday Book: an extensive population count and property register commissioned by William the Conqueror in 1066 (505-506).

Malthus: argued that population grows faster than the food supply. Population growth is more likely to be checked by an increase in mortality (514-516).

Davis: developed the theory of the demographic transition. Theory traces a society's transition from a high fertility - high mortality environment to a low fertility - low mortality environment. Population growth occurs when the mortality rate falls before the fertility rate. Persons choose to limit fertility once large families are perceived as costly (519-520).

Berelson: identified seven modernization "thresholds" that are associated with fertility decline. These thresholds involve declining agricultural labor force participation, increased educational involvement and literacy, increased life expectancy, a decline in infant mortality, a delay in the age at first marriage and improved standard of living. Fertility decline occurs when several "thresholds" are attained (520)

Cutright and Hargens: studied fertility decline in Latin America. Fertility declines occurred when literacy and life expectancy thresholds were attained (520).

Bongaarts: observed that actual fertility is significantly higher than wanted fertility in high fertility nations (529).

Caldwell and colleagues: research suggests that unwanted births are high in developing countries because contraceptives cannot be obtained easily (529-530).

Info Trac Search Words

Enter these search terms to conduct more extensive investigations of key topics introduced in Chapter 18.

Population Change

Select a study that interests you. Does your study focus primarily on changes in fertility, mortality or migration? Are any population projections offered?

Malthus, Thomas Robert

Demographic Transition

Subdivisions: Forecasts

Epidemiological Transition

Baby Boom Generation

Subdivisions: Social Aspects

Select one of the studies listed. Note how the baby boom generation has influenced U.S. society. Are any future impacts discussed?

Multiple Choice

Answers and page references are provided at the chapter end.

1. A total count of a population is a
   a. sample.
   b. cohort.
   c. sex ratio.
   d. growth rate.
   e. census.
2. The study of human population and population characteristics is
   a. geography.
   b. demography.
   c. anthropology.
   d. theology.
   e. ethnography.

3. The three components of population growth are
   a. births, deaths, and migration.
   b. expansive, stationary and expansive age structures.
   c. race, class and gender.
   d. positive, negative and preventive checks.
   e. rates, cohorts and population structures.

4. The number of births per 1,000 population is the
   a. age-specific fertility rate.
   b. reproductive age rate.
   c. crude birth rate.
   d. total fertility rate.
   e. vital statistics rate.

5. In 1979 the average Kenyan woman gave birth to 8.1 children during her reproductive age years. This is an example of a/an
   a. crude birth rate.
   b. birth cohort.
   c. positive check.
   d. fertility rate.
   e. zero population growth rate.

6. All the persons born in the U.S. in the year 2000 represent a
   a. constrictive age structure.
   b. expansive sex structure.
   c. baby boom.
   d. birth cohort.
   e. geometric growth rate.

7. Younger cohorts are larger than older cohorts in a/an
   a. exponential population structures.
   b. stationary population structures.
   c. expansive population structures.
   d. constrictive population structures.
   e. arithmetic population structures.

8. A major source of decline in population growth among preindustrial societies is
   a. famine.
   b. disease.
   c. war.
   d. all of the above
   e. none of the above

9. Famine and plague reduced Europe's population __________ percent between 1300-1400 A.D.
   a. 40%
   b. 25%
   c. 15%
   d. 10%
   e. 5%

10. The *Essay on the Principle of Population* was written in 1798 by
   a. Marx.
   b. Smith.
   c. Davis.
   d. Weber.
   e. Malthus.

11. Which of the following statements is false?
   a. Malthus' theory looks at the association between population growth and the food supply.
   b. Population growth is exponential.
   c. The food supply grows exponentially.
   d. Famine and disease are positive checks on population growth.
   e. Fertility rates will remain high and not restrict population growth.

12. During the early stages of the Industrial Revolution, many European nations experienced an increase in population growth. This increase was primarily due to a/an
   a. increase in mortality.
   b. decrease in mortality.
   c. decrease in migration.
   d. increase in migration.
   e. decrease in fertility.

13. Replacement-level fertility occurs when the average woman gives birth to slightly more than
    a. one child.
    b. two children.
    c. three children.
    d. four children.
    e. five children.

14. The demographic transition involves the transition from
    a. industry to agriculture.
    b. urban life to rural life.
    c. a high fertility and mortality environment to a low fertility and mortality environment.
    d. expansive population structures to constrictive population structures.
    e. high rates of migration to low rates of migration.

15. Davis argued that fertility declines because
    a. small families are perceived as costly.
    b. large families are viewed as costly.
    c. contraceptives became more readily available.
    d. religious sanctions against birth control no longer exist.
    e. governments establish coercive population control policies.

16. According to Davis the initial outcome of the modernization process is
    a. a sudden decline in fertility.
    b. an increase in urban to rural migration.
    c. the transition from a skilled labor force to an unskilled labor force.
    d. a sudden decline in mortality.
    e. a sudden decline in migration.

17. Each of the following is a modernization threshold identified by Berelson except
    a. average life expectancy reaches 60 years.
    b. infant mortality falls to 65 deaths per 1,000 infants.
    c. per capita GNP reaches $450.
    d. adult literacy rates reach 70%.
    e. the sex ratio reaches 110.

18. According to the *Nations of the Globe* data, the strongest negative correlation is between the fertility rate and
    a. high school enrollment.
    b. the infant mortality rate.
    c. economic development.
    d. literacy.
    e. percentage of persons not employed in agriculture.

19. Cutright and Hargens observed that the fertility decline in Latin American countries was linked to meeting these two modernization thresholds.
    a. literacy and percent of the population not employed in agriculture
    b. the infant mortality rate and per capita GNP
    c. life expectancy and school enrollment
    d. literacy and life expectancy
    e. early age at first marriage and per capita GNP

20. The baby boom cohort includes persons born between
    a. WWI and WWII.
    b. the Great Depression and WWII.
    c. 1946 and the mid-1960s.
    d. the decade of the 1970s.
    e. 1985 to 1999.

21. The baby boom accompanied a massive
    a. migration to the Northeast.
    b. migration to the West Coast.
    c. expansion of urban inner-city population.
    d. expansion of the suburbs.
    e. expansion of rural villages.

22. Demographic transition theory suggests that the long-term effect of modernization is
    a. exponential population growth.
    b. exponential growth in the food supply.
    c. population stability.
    d. a higher sex ratio.
    e. a lower sex ratio.

23. The demographic transition in developing nations
    a. has been faster then the European transition.
    b. has been at about the same pace as the European transition.
    c. has been slower than the European transition.
    d. has not yet started.
    e. none of the above

24. Over the 1965-2001 period, fertility rates for many Asian, Latin American and sub-Saharan African nations have
    a. remained the same.
    b. declined.
    c. increased.
    d. increased only for women in polygamous marriages.
    e. declined only for women in second marriages.

25. China's fertility decline has been influenced by
   a. an increase in standard of living.
   b. a decrease in standard of living.
   c. a coercive government-sponsored population policy.
   d. a and c
   e. b and c

26. Measures of ideal family size are examples of
   a. baby boom cohorts.
   b. population structures.
   c. sex structures.
   d. sex ratios.
   e. numeracy about children.

27. Bongaarts has observed that in many sub-Saharan African nations
   a. actual fertility equals wanted fertility.
   b. wanted fertility is lower than actual fertility.
   c. wanted fertility is higher than actual fertility.
   d. the fertility rate is below replacement level.
   e. female children are preferred over male children.

28. The potential for depopulation presently is the greatest in this region of the world.
   a. Latin America
   b. India
   c. South America
   d. sub-Saharan Africa
   e. Europe

29. Rapidly declining fertility rates in industrial nations suggest that couples may be
   a. placing more value on their leisure time.
   b. placing more value on attaining an affluent lifestyle.
   c. viewing children as interfering with their freedom.
   d. b and c.
   e. a, b and c.

30. Current population projections indicate that by 2050 approximately __________ of Europeans will be over age sixty.
   a. 10 %
   b. 20 %
   c. 33 %
   d. 50 %
   e. 67 %

True/False

Answers and page references are provided at the chapter end.

1. The *Domesday Book* included a count of English population and property ownership around 1066.

2. The crude birth rate is calculated by dividing the total number of births by the population of women of reproductive age.

3. Constrictive population structures are associated with declining population growth.

4. Three major factors contributing to low mortality rates in preindustrial societies are famine, disease and war.

5. Malthus argued that population grows at an exponential rate.

6. Malthus identified fertility control as an example of a positive check on population growth.

7. When a society experiences replacement-level fertility, births equal deaths.

8. According to the demographic transition theory, high population growth rates are experienced when fertility rates fall below mortality rates.

9. The demographic transition in developing countries has been slow and gradual.

10. Caldwell and colleagues maintain that unwanted births in many developing countries is high because many people experience difficulty in obtaining contraceptives.

Short Answer Questions

These short answer questions are provided to test your knowledge and understanding of the basic sociological concepts presented in Chapter 18. Page references for answers are included at the chapter end.

1. Why do many demographers consider the fertility rate to be a more valid measure of fertility?

2. Identify the three primary population structures.

3. According to Malthus what is the nature of the association between population growth and the food supply?

4. How did the mechanization of agriculture and modernization contribute to the decline in mortality in Europe.

5. What is replacement-level fertility?

6. Describe the major tenets of Davis' demographic transition theory.

7. Identify Berelson's modernization thresholds and state their relationship to fertility reduction.

8. What is the baby boom, and how has this cohort influenced U.S. life?

9. Evaluate the association between economic development and fertility reduction utilizing China as a test case.

10. Identify several potential problems associated with "depopulation."

Essay Questions

These questions are designed to test your understanding of key sociological concepts presented in Chapter 18 and your ability to apply these insights to concrete situations.

1. Censuses can be very costly particularly when response rates are low. Identify several key factors that would influence response rates and note some possible ways to reduce census costs.

2. The U.S. baby boomers are a well-known birth cohort. How will the baby boomers impact the quality of U.S. life once they retire, and how will their retirement be financed.

3. Explain why population growth, food supply, economic development and public health are interdependent phenomena.

4. The theory of the demographic transition has become a classic demographic theory. Do nations that enter the transition necessarily complete the transition? Is it possible to become trapped in the high population growth stage?

5. Fertility surveys in developing nations are now beginning to include data on male contraceptive use and male attitudes toward fertility. Why is this important?

Answers

Multiple Choice

1. e (506)
2. b (506-507)
3. a (507)
4. c (508)
5. d (508)
6. d (509)
7. c (510-511)
8. d (512-514)
9. a (514)
10. e (515)
11. c (514-517)
12. b (517)
13. b (518)
14. c (518)
15. b (519)
16. d (520)
17. e (520)
18. b (520)
19. d (520)
20. c (521)
21. d (521-522)
22. c (525)
23. a (520, 525)
24. b (526)
25. d (527-528)
26. e (528)
27. b (529)
28. e (530-532)
29. e (531)
30. d (532)

True/False

1. T (505-506)
2. F (508)
3. T (512)
4. F (512)
5. T (515)
6. F (515-516)
7. T (518)
8. F (519-520)
9. F (520, 524-525)
10. T (529-530)

Short Answer

1. (508)
2. (510-512)
3. (514-516)
4. (517)
5. (518, 530)
6. (519-520)
7. (520, 528)
8. (521-523)
9. (527-528)
10. (530-532)

CHAPTER 19

URBANIZATION

Extended Abstract

An increasing number of people throughout the world are residing in cities and not relying on farming for work; however, urban societies are a recent phenomenon. It is believed that Great Britain was the world's first urban society, and this was not attained until the twentieth century. In this chapter Stark reviews life in preindustrial cities, examines the association among agricultural development, transportation networks and urban development, notes the growth of suburbs, addresses commuting and urban segregation and finally comments on the quality of urban life.

Preindustrial cities were small in terms of population size, but they were densely populated. Transportation networks were poor and food could not be transported easily. Infectious disease rates were high and sanitation was poor. Pollution from wood-burning and coal-burning homes and businesses was problematic, and personal hygiene was poor as persons rarely bathed. Since many city streets were not lighted at night, crime was high. Yet, even with these problems, people still migrated to the cities. Cities offered economic opportunities, cultural stimulation, tolerated more risky behaviors and enabled persons to become more anonymous.

Industrialization and urbanization are strongly correlated, but each is contingent upon agricultural development and the development of more extensive transportation networks. Increased agricultural productivity accompanied the mechanization of agriculture. In 1820 each U.S. farm worker could supply enough food to support 4 persons, but by 1970 each U.S. farm worker was supporting 47 people! As farmers became more productive, more people were free to move to cities in search of economic opportunities. Industrialization also promoted increased job specialization and occupational interdependence.

An area is identified as an urban place once its population reaches 2,500 persons. The U.S. Census classifies areas as cities once the population reaches 50,000 inhabitants. As cities expand, suburbs and urban centers form. These cities and adjacent suburbs form metropolitan areas. Within these metropolitan areas, job ties, leisure activities, mass communication networks and a sense of identity link city and suburban residents. Clearly defined metropolitan areas are labeled Standard Metropolitan Statistical Areas (SMSAs) in the U.S. and Census Metropolitan Areas (CMAs) in Canada.

The other major factor impacting urban growth is the development of more complex transportation networks. Two classic urban transportation networks are the fixed-rail metropolis and the freeway metropolis. Railroad and trolley routes determine the course of urban development in the fixed-rail metropolis. Within this urban development model, industries concentrate in the center of the city. Since workers need to be close to their work and retailers need to be close to residential areas, fixed-rail transportation networks extend from the city center linking industry, retailers and workers. Areas adjacent to the rail lines and the city center become high-density areas. These networks changed as urban dwellers began to rely more on cars and trucks to meet transportation needs. Road systems are more cost efficient and reduce the need for activities to be concentrated in the city center and along rail lines. Industrial development and shopping centers could now be transferred to the suburbs as the freeway metropolis fostered industrial, retail and residential decentralization. Concerns over proximity were replaced with concerns over traffic flow and travel time.

U.S. residents appear to prefer life in decentralized, lower density urban environments. This is evident in the growth of the suburbs, the location of businesses and industrial establishments in lower density areas and the avoidance of public transportation. According to the 2000 census, the average commute time from home to work is 25.5 minutes, and three out of four workers commute to work alone using their own vehicle. The law of constant travel time states that people will move if their round trip commute exceeds one hour. Finally, Gallup Poll data indicate that only 13% of U.S. residents prefer to live in a city, and among persons presently living in cities, only 20 % preferred living in cities. Presently, over half of the U.S. population resides in suburbs.

Stark next turns to a discussion of urban neighborhood segregation and integration. Urban neighborhoods tend to be stratified by social class and by racial and ethnic background. Park and Burgess offered a classical explanation of racial and ethnic neighborhood segregation linking neighborhood segregation and status inequality. According to their theory of ethnic succession, lower status immigrant groups are segregated in poorer neighborhoods. As immigrant groups experience economic success, they move into more integrated, higher status neighborhoods. Poor neighborhoods are therefore continuously occupied by a series of lower status groups. On the other hand, Darroch and Marston maintain that neighborhood segregation persists even when status inequality among individuals is reduced. Kantrowitz reports similar findings.

Responding to these claims, Guest and Weed analyzed data on racial and ethnic neighborhood segregation in three racially and ethnically diverse cities. Employing the Taeubers' index of dissimilarity as a measure of neighborhood integration, Guest and Weed observed that neighborhoods became more integrated during the 1960s. These findings suggest that racial and ethnic neighborhood segregation declines as between group status inequality disappears. Recent census data indicate that segregation in major U.S. cities is declining and that segregation is least pronounced in the newer cities of the West and South.

The impact of urbanization on quality of life is addressed in the last section. In a classic distinction of rural and urban life, Tonnies maintained that *Gemeinschaft* communities are small, close-knit societies characterized by a high degree of conformity to the norms. Communities function as primary groups. *Gesellschaft* communities are larger and impersonal. People are motivated by self-interest; relationships are more manipulative, and there is less agreement with the norms. Likewise, Durkheim argued that since urbanization weakens the power of norms to regulate social behavior and strengthen attachments (anomie), deviance will be higher in urban societies.

In another classic study on urban life, Wirth claimed that urban dwellers are more likely to isolate themselves. Adopting a similar perspective, Faris and Dunham link urbanization and social disorganization and argue that social disorganization contributes to a higher incidence of mental illness. Furthermore, disorganization would be more pronounced in poorer neighborhoods. Subsequent research has demonstrated that mental illness is not an outcome of urban social disorganization and poverty. Persons experiencing mental problems are more likely to "drift" into poorer neighborhoods as a result of unemployment.

Researchers have also argued that urban crowding or living in high-density areas increases the incidence of certain social pathologies like alcoholism. Researchers have been unable to clearly demonstrate that density impacts behavior in a negative manner at the community level. Micro level studies of crowding suggest that persons residing in crowded homes are more likely to withdraw from others; however, the percentage of persons residing in crowded homes appears to be declining.

Key Learning Objectives

After a thorough reading of Chapter 19, you should be able to:

1. identify important distinguishing characteristics of preindustrial cities and describe the quality of life in preindustrial cities.

2. indicate why the mechanization of agriculture and urbanization are interrelated.

3. note important advances in agricultural productivity.

4. differentiate urban place, city, metropolitan area and SMSA.

5. distinguish the fixed-rail and freeway metropolis.

6. cite evidence indicating that people prefer living in decentralized cities.

7. state the association between the law of constant travel time and a person's willingness to relocate.

8. indicate why minority groups may concentrate in certain neighborhoods.

9. critically evaluate the Park and Burgess theory of ethnic succession.

10. describe how the index of dissimilarity is calculated and interpreted.

11. document trends in urban segregation and desegregation over the past forty years.

12. distinguish *Gemeinschaft* and *Gesellschaft* communities.

13. evaluate ecological theories linking urbanization, social disorganization and social pathologies.

14. articulate the "social drift" explanation for urban social pathologies.

15. assess the impact of urban density and crowding on quality of life.

Chapter Outline

I. An Overview of Preindustrial Cities
   A. Small and Densely Populated
   B. Poor Transportation Networks and the Food Supply

C. Centers of Disease and Pollution
D. Limited Crime Control
E. The Attraction of Cities
    1. Cities were a source of economic opportunity.
    2. Cities were centers of culture and "vice."

II. Agricultural Development, Transportation Networks and Urbanization
A. Mechanization of Agriculture and Productivity
    1. Technology is employed to increase agricultural productivity.
    2. Division of labor expands, as fewer agricultural workers are needed.
    3. Occupational specialties locate in urban centers.
B. Measures Assessing Degree of Urbanization
    1. Urban places include locations with populations of 2,500 or more.
    2. Cities maintain a population of at least 50,000 residents.
    3. Suburbs are small communities adjacent to cities.
    4. A metropolitan area includes a city and any adjacent suburbs.
    5. Standard Metropolitan Statistical Areas (SMSAs) and Census Metropolitan Areas (CMAs) represent major urban areas that are linked through labor force ties.
C. Transportation Networks and Urban Growth
    1. The fixed-rail metropolis exemplifies centralized urban development.
        a. Fixed-rail transportation networks extend from the city center linking industry, retailers and workers.
        b. Areas adjacent to the rail lines and the city center are high- density areas.
    2. The freeway metropolis exemplifies decentralized urban development.
        a. Road systems are more cost efficient and reduce the need for activities to concentrate in the city center and along rail lines.
        b. Concern over proximity of activities is replaced by a concern over traffic flow and travel time.
    3. The average commute time for U.S. workers is 25.5 minutes, and three out of four workers travel to work alone in their own vehicles.
    4. According to the law of constant travel time, people will move if their round trip commute time to work exceeds one hour.
D. Growth of Suburbs
    1. U.S. residents appear to prefer life in decentralized, low density urban environments.
    2. Business and industrial establishments have moved to low density areas.

III. Urban Neighborhood Segregation
A. The Theory of Ethnic Succession: Park and Burgess
    1. Neighborhood segregation is linked to status inequality.
    2. Lower status immigrant groups are segregated in poorer neighborhoods.
    3. Immigrant groups move into more integrated higher status neighborhoods as economic success is experienced.
    4. Poorer neighborhoods are continuously occupied by a series of lower

status groups.

B. Ethnic Succession: Conflicting Findings
   1. Darroch and Marston's study indicates that neighborhood segregation persists even when status inequality among individuals is reduced.
   2. Guest and Weed's study demonstrates that racial and ethnic neighborhood segregation declines as between group status inequality disappears.
   3. Recent census data suggest that segregation in major U.S. cities is declining and that newly built cities in the West and South are the least segregated.
   4. The ten most segregated U.S. cities are the old industrial cities.

IV. Urbanization and Quality of Life
   A. *Gemeinschaft* and *Gesellschaft* Communities (Toennies)
      1. *Gemeinschaft* communities are small, close-knit societies characterized by a high degree of conformity to the norms.
      2. Persons residing in *Gesellschaft* communities are motivated by self-interest, and there is less agreement on the norms.
   B. Urbanization, Social Disorganization and Social Pathologies
      1. Durkheim maintained that since urbanization weakens the norms (anomie), deviance is higher in urban societies.
      2. Wirth claimed that urban dwellers are more likely to isolate themselves.
      3. Faris and Dunham link urbanization, social disorganization and higher incidence of mental illness. Social disorganization is more pronounced in poorer neighborhoods.
      4. Subsequent research challenges findings of Faris and Dunham.
         a. Mental illness is not an outcome of urban social disorganization and poverty.
         b. Persons experiencing mental problems are more likely to "drift" into poorer neighborhoods as a result of unemployment.
   C. Urban Crowding, Density and Social Pathologies
      1. There is no clear evidence that density impacts behavior in a negative manner at the community level.
      2. Micro level studies of crowding suggest that persons living in crowded homes are more likely to isolate themselves from others.
      3. The percentage of persons residing in crowded homes appears to be declining.

Key Terms

Based on your reading of Chapter 19, you should be able to define and illustrate the key sociological concepts listed below. Page numbers are provided in parentheses as reference points.

Urban Society (539)
Urbanization (539)
Specialization (549)
Urban Place (550)
City (550)
Suburb (550)

Metropolitan Area (550)
Metropolis (550)
Sphere of Influence (550)
Standard Metropolitan Statistical Area [SMSA] (550)
Census Metropolitan Area [CMA] (551)
Fixed-Rail Metropolis (551)
Freeway Metropolis (551)
Law of Constant Travel Time (555)
Theory of Ethnic Succession (557)
Index of Dissimilarity (557)
Gemeinschaft (560)
Gesellschaft (560)
Anomie (560)
Urbanism as a Way of Life (561)
"Social Drift" Explanations of Urban Social Pathologies (562)
Mass Society Theory (562)

Key Research Studies

Listed below are key research studies cited in Chapter 19. Familiarize yourself with the major findings of these studies. Page references are provided in parentheses.

Graunt: published the first set of vital statistics for London in 1662. The data were based on weekly death reports for each London parish (544).

Zahavi: formulated the "law of constant travel time." Persons will move if the round trip work commutes exceeds one hour (555).

Park and Burgess: developed the theory of ethnic succession. According to this theory, lower status groups are segregated in poorer neighborhoods. As lower status groups gain economic success, they move out of poorer neighborhoods and are replaced by other lower status groups (556-557).

Darroch and Marston; Kantrowitz: research indicates that neighborhood segregation persists even when status inequality among groups is reduced (557).

Taeuber and Taeuber: developed the index of dissimilarity to measure neighborhood segregation. The measure compares the racial-ethnic composition in a city block to the racial-ethnic composition of the city. Higher scores signal a greater degree of neighborhood segregation (557-558).

Guest and Weed: used index of dissimilarity to measure degree of integration achieved within three diverse SMSAs. Neighborhoods in each SMSA became more integrated over the 1960-1970 time period. Neighborhoods segregated on the basis of race and ethnicity decline as status inequality declines (557-558).

Faris and Dunham: attempted to link urbanization, social disorganization and mental illness. They argued that social disorganization is more pronounced in poorer neighborhoods and portrayed mental illness as an outcome of urban social disorganization and poverty. The interpretation of these findings is challenged by the "social drift" argument, which maintains that persons experiencing mental illness are more likely to move into poorer neighborhoods as a result of unemployment (572-573).

Gove, Hughes and Galle: conducted a micro level study assessing the impact of crowding on behavior. They observed that persons living in crowded homes are more likely to isolate themselves from others (564).

Info Trac Words

Enter these search terms to conduct more extensive investigations of key topics introduced in Chapter 19.

Urbanization

Subdivisions: Environmental Aspects

Select one of the articles listed. How is urban growth impacting the environment? Are any solutions or policy recommendations offered?

City and Town Life

Subdivisions: Health Aspects

Several interesting activities are identified in this search. Select one. How healthy are contemporary cities? How do the health conditions described in your selected study compare with the conditions associated with preindustrial cities? Are contemporary cities a better place to live?

Gridlock

This is a transportation studies term. Take a look at one of the articles that addresses gridlock as a transportation problem.

Index of Dissimilarity

Urban Density

Multiple Choice

Answers and page references are provided at the chapter end.

1. The first urban society in history was
   a. China.
   b. India.
   c. the United States.
   d. Great Britain.
   e. France.

2. Presently, __________ percent of U.S. residents live in urban areas and ____________ percent live on farms.
   a. 77; 23
   b. 23; 77
   c. 23; 2
   d. 79; 1
   e. 50; 50

3. Preindustrial cities were
   a. small.
   b. unhealthy.
   c. crowded.
   d. dangerous at night.
   e. all of the above

4. The typical preindustrial city had a population of __________ inhabitants.
   a. 5,000 to 10,000
   b. 25,000 to 50,000
   c. 50,000 to 75,000
   d. 100,000 to 250,000
   e. 500,000 to 1,000,000

5. Mortality rates in preindustrial cities were high. This is because
   a. cities were heavily polluted.
   b. persons were living in crowded areas and were exposed to infectious diseases.
   c. the water supply was often contaminated.
   d. food spoiled easily.
   e. all of the above

6. Which of the following statements about life in preindustrial cities is false?
   a. Since the cities were so filthy, persons bathed often.
   b. Human waste was often dumped into trenches running along the streets.
   c. Home fireplaces were a major source of pollution.
   d. People often wore high boots because the streets were littered with manure and garbage.
   e. The water quality was often poor and not safe to use.

7. Preindustrial cities tended to be characterized by
   a. large population size and high population density.
   b. small population size and low population density.
   c. small population size and high population density.
   d. large population size and low population density.
   e. a larger older population and a surplus of females.

8. Preindustrial cities attracted persons who
   a. were in search of economic opportunity.
   b. wanted to enjoy more cultural opportunities.
   c. wanted to lead a more anonymous life.
   d. were adventuresome, single and young.
   e. all of the above

9. From 1820 to 1970 the number of U.S. residents being supplied with farm products by one U.S. worker increased 4 persons to __________ persons.
   a. 10
   b. 15
   c. 26
   d. 47
   e. 75

10. From 1800 to 1980 the number of hours of labor required to harvest an acre of wheat has
   a. increased, and the yield per acre has increased.
   b. decreased, and the yield per acre has decreased.
   c. decreased, and the yield per acre has increased.
   d. increased, and the yield per acre has decreased.
   e. remained the same, but the yield per acre has increased.

11. An elaborate division of labor is known as
   a. urbanization.
   b. industrialization.
   c. specialization.
   d. transportation.
   e. integration.

12. To be classified as a city by the U.S. census, a location must have at least _________ residents.
   a. 2,500
   b. 5,000
   c. 10,000
   d. 25,000
   e. 50,000

13. An urban place is any location with at least _________ inhabitants.
   a. 2,500
   b. 2,000
   c. 1,500
   d. 1,000
   e. 500

14. This geographic area presently has the largest percentage of persons living in urban places.
   a. the United States
   b. Hong Kong
   c. Belgium
   d. China
   e. India

15. A small community in the immediate vicinity of a city is generally labeled a
   a. SMSA
   b. CMA
   c. rural village
   d. suburb
   e. metropolis

16. Persons residing near a city are often dependent upon the city for the provisions of jobs, recreational opportunities and a feeling of community identity. This is known as the city's
   a. sphere of influence.
   b. range of anomie.
   c. urban place.
   d. index of dissimilarity.
   e. index of similarity.

17. SMSA is an abbreviation for
   a. Statically Monitored Social Area.
   b. Social Measurement Statistical Area.
   c. Standard Metropolitan Statistical Area.
   d. Standard Mexican Statistical Area.
   e. Standard Measurement Stastical Area.

18. A SMSA includes
   a. a city with at least 50,00 inhabitants.
   b. counties with at least 70% of the work force employed in agriculture.
   c. counties with at least 75% of the work force employed outside of agriculture.
   d. a and b
   e. a and c

19. Each of the following is an identifying feature of the fixed-rail metropolis except
   a. trolleys were introduced as an important mode of transportation.
   b. the center of the city was the focal point of industrial and cultural activity.
   c. cities began to develop outward.
   d. cars and trucks replaced trolleys as the main mode of transportation.
   e. metropolitan expansion was tied to the expansion of fixed-rail lines.

20. In the freeway metropolis
   a. roads replaced rail lines.
   b. cities became more decentralized.
   c. shopping malls replaced downtown shopping.
   d. most commercial goods were now transported by trucks.
   e. all of the above

21. According to the 2000 census, the average travel time from home to work is
   a. 10.0 minutes.
   b. 15.7 minutes.
   c. 20.3 minutes.
   d. 25.5 minutes.
   e. 30.0 minutes.

22. The greatest percentage of U.S. residents presently live
   a. on farms.
   b. in suburban communities.
   c. in small rural villages.
   d. in the downtown areas of major cities.
   e. in cities with a population of 250,000 or more.

23. According to Park and Burgess, slum neighborhoods are
   a. successively occupied by the lowest status groups at that time.
   b. successively occupied by the highest status groups at that time.
   c. not segregated by social class or social status.
   d. not segregated by race or ethnicity.
   e. becoming less crowded.

24. In testing the claims of the theory of ethnic succession, Guess and Weed observed that
   a. ethnic groups chose to continue to live in segregated neighborhoods even when status equality had been obtained.
   b. individual measures of status attainment are more important than a group's overall level of status attainment.
   c. neighborhoods become more integrated as between group status inequality is reduced.
   d. neighborhoods become more integrated as between group status inequality increases.
   e. U.S. neighborhoods have become more segregated over time.

25. A standard measure of degree of segregation and integration is the
   a. index of dissimilarity.
   b. index of similarity.
   c. sphere of influence.
   d. SMSA.
   e. CMA.

26. From 1990 to 2000, neighborhood segregation in many of the larger U.S. metropolitan areas has
    a. increased.
    b. remained the same.
    c. declined.
    d. caused commute times to increase.
    e. caused commute times to decrease.

27. According to 2000 census data, __________ is the most segregated U.S. metropolitan area.
    a. Chicago
    b. Detroit
    c. Las Vegas
    d. Honolulu
    e. Cleveland

28. Each of the following is a characteristic of *Gemeinschaft* life except
    a. persons are motivated by self-interest.
    b. small, cohesive communities.
    c. friendship and kinship bonds are strong.
    d. the community functions as a primary group.
    e. people agree on the norms.

29. Durkhein associated urbanization with
    a. higher levels of deviance.
    b. anomie.
    c. the weakening of social attachments.
    d. social disorganization.
    e. all of the above

30. Faris and Dunham tried to demonstrate that
    a. crowding in urban areas is declining.
    b. mental illness is an outcome of urbanization, poverty and social disorganization.
    c. persons suffering from mental illness tend to "drift" into poor neighborhoods as a result of unemployment.
    d. crowding in urban areas is increasing.
    e. social attachments are stronger in urban areas than rural areas

True/False

Answer and page references are provided at the chapter end.

1. Urbanization involves the migration of people from the cities to rural communities and villages.

2. Preindustrial cities were large, safe, clean and healthy.

3. Graunt published the first set of vital statistics for the city of London in 1662.

4. Urbanization is contingent upon increased agricultural productivity and the development of adequate transportation networks.

5. To be classified as an urban place, a location must have a population of 2,500.

6. Railroads and trolley lines are the identifying features of the freeway metropolis.

7. Presently over half of the U.S. population lives in a rural villages.

8. Park and Burgess linked racial and ethnic neighborhood segregation to status differences among groups.

9. *Gemeinschaft* is to *Gesellschaft* as urban rural village is to urban metropolis.

10. According to the "social drift" explanation of mental illness, the social disorganization characteristic of urban neighborhoods promotes mental illness.

Short Answer Questions

These short answer questions are provided to test your knowledge and understanding of the basic sociological concepts presented in Chapter 19. Page references for answers are included at the chapter end.

1. Why were mortality rates in preindustrial cities high?

2. Provide examples illustrating how technological advancements have increased agricultural productivity.

3. Distinguish urban place, city, and suburb.

4. What criteria must be met if an urban area is to be classified as a Standard Metropolitan Statistical Area?

5. Identify the major modes of transportation associated with the fixed-rail metropolis and the freeway metropolis.

6. State four reasons why people appear to prefer living in decentralized cities.

7. What is the "law of constant travel time?"

8. State the major tenets of the Park and Burgess theory of ethnic succession.

9. What is the index of dissimilarity and how is it interpreted?

10. How does the "social drift" explanation account for the overrepresentation of the mentally ill in poorer neighborhoods?

Essay Questions

These questions are designed to test your understanding of key sociological concepts presented in Chapter 19 and your ability to apply these insight to concrete situations.

1. Compare the health conditions experienced in the urban areas of developing nations to the health conditions experienced in preindustrial European cities. Is the quality life in these contemporary urban areas better, worse or essentially the same?

2. How is agribusiness impacting the association among urbanization, agricultural development and the development of transportation networks?

3. In many colleges and universities, students may take classes in urban sociology and rural sociology. Outline major topics that would be addressed in a course on the sociology of the suburbs.

4. Racial and ethnic diversity is increasing in the U.S. Will this lead to an increase or decrease in neighborhood segregation? Justify your answer.

5. Urban planners, residents, developers and municipal leaders often are involved in heated debates over the impact of density on quality of urban life. Why is density such a controversial issue?

Answers

Multiple Choice

1. d (539)
2. d (539)
3. e (540)
4. a (540)
5. e (541-543)
6. a (542)
7. c (540, 543)
8. e (544-546)
9. d (546-547)
10. c (547, 549)
11. c (549)
12. e (550)
13. a (550)
14. b (550)
15. d (550)
16. a (550)
17. c (550)
18. e (550-551)
19. d (552-553)
20. e (552-553)
21 d (554-555)
22. b (556)
23. a (557)
24. c (557-558)
25 a (557)
26. c (558-559)
27. b (559)
28. a (560)
29. e (560)
30. b (561)

True/False

1. F (539)
2. F (540-544)
3. T (544)
4. T (546-549, 551-553)
5. T (550)
6. F (552-553)
7. F (556)
8. T (556-557)
9. T (560)
10. F (561-562)

Short Answer

1. (542-544)
2. (546-549)
3. (550)
4. (550-551)
5. (552-553)
6. (553-554)
7. (555)
8. (556-557)
9. (557-558)
10. (562)

CHAPTER 20

THE ORGANIZATIONAL AGE

Extended Abstract

As a group's size increases, leaders encounter more difficulties in managing the groups' affairs. Effective leaders try to maintain control over large, more complex groups, also known as formal organizations, by stating goals, establishing procedures, selecting and training support staff, identifying a chain of command and maintaining records to gauge progress. These large, formal organizations are rational organizations. In this chapter the rise of formal organizations is reviewed, Weber's model of bureaucracy is presented, rational and natural systems are evaluated and problems associated with diversification and decentralization are addressed.

Formal organizations are recent phenomena having originated in the nineteenth century. The early growth of formal organizations was particularly evident within the military, business and government service. By the beginning of the nineteenth century, individual armies had growth to the point that one person could not lead them effectively. Noting this problem, Moltke, a Prussian Field Marshal, decided to select and train a military staff that would carry out his objectives. Since this staff received the same training, they were interchangeable and could be placed in charge of smaller standardized military units, which comprised a divisional system. By dividing the army into smaller, standardized units and delegating responsibility, Moltke was able to regain control of a large formal organization, the army.

In the field of business, the accomplishment of McCallum and Swift are worth noting. McCallum observed that short-distance rail lines were more efficient than long-distance rail lines. To enhance the efficiency of the longer lines, McCallum created many short-distance links within one long-distance run. A supervisor who would report to the main office controlled each short-distance link, a geographical division. Again, control of the more complex organizations was gained through the delegation of authority and the coordination of activities. In a similar manner, Swift discovered that control over the meat-packing industry could be obtained if the activities of the organization were coordinated from beginning to end. Managers were therefore placed in charge of buying, packing, shipping and marketing. Thus, Swift created functional divisions and utilized the principle of vertical integration to maintain control of the company.

The civil service system was introduced as a means of combating the disorganization and inefficiency associated with the spoils system. Under the spoils system, elected government officials could reward their supporters with government appointments. This created an unstable organization since administrations could be replaced with each election and incompetent people could be rewarded. Under the civil service system, government workers are hired and promoted on the basis of technical qualifications and merit.

Weber offers the classic study of formal organizations. Formal organizations were conceptualized as bureaucracies, organizations whose activities are based on the development of rational rules and procedures. These rules and procedures insure that the actions of the bureaucracy are predictable and controlled. Thus, bureaucratic organizations are comprised of many areas of functional specialization. The activities of these areas of specialization are specified in formal sets of rules and practices. Carefully trained leaders and managers insure that the goals and objectives of the organization are met since power and authority are maintained through a hierarchical chain of command. The progress and success of the members of the bureaucracy are monitored through an extensive record keeping system. Weber argued that bureaucracies developed as a means of enhancing the efficiency of larger formal organizations. But, do bureaucracies really behave in this manner? What about internal conflict, inefficiency and failure to comply with an organization's goals and objectives? In addressing these issues, contemporary researchers distinguish rational and natural systems.

According to the rational system approach, bureaucracies are goal-oriented organizations. They exist to satisfy a need in an efficient manner; however, bureaucracies do not always function as intended. Some goals are reinstated, and all members of the organization may not pursue the same goals. Alternative communication and power networks may characterize the organization. These concerns are addressed in a natural system approach to formal organization. Natural system approaches emphasize such issues as goal displacement, goal conflict and informal relations. Survival is an unwritten goal in many organizations. When an organization's continued existence is threatened, stated goals may be replaced to enhance survival. For example, once a cure for polio was discovered, the focus of The National Foundation for the March of Dimes shifted from polio to birth defects. Goal conflict is illustrated by the fact that labor and management often pursue different goals within the same organization. Furthermore, within a formal organization, persons are often able to circumvent the chain of command. Consequently, rational and natural system approaches to formal organizational behavior are complementary approaches.

Functional divisions represent an effective approach to maintaining control over formal organizations that pursue one major task. However, when organizations pursue multiple tasks, functional divisions can be disastrous. For example, prior to WWI, DuPont was a major producer of gunpowder and explosives. After the war, DuPont explored and developed other markets rather than scale back production. Management quickly discovered that functional divisions designed to coordinate the production of one major product were very ineffective when multiple products were involved. Since the needs of each product line are different, the needs of the company could be served best if each product division were treated as a separate business and a functional divisional system was implemented for each product. Control of the larger organization was thus maintained through the creation of autonomous divisions and through decentralization.

In developing a theory of administrative growth, Blau argues that the cost of management increases as an organization increases in size and becomes involved in more diverse activities. Since smaller organizations devote a smaller percentage of their operating budget to management costs, Blau's theory suggests that decentralized organizations may be more efficient.

As organizations continue to grow, top management is increasingly more removed from the direct supervision of many organizational activities. This has led to a need to delegate decision-making authority (discretion) to subordinates who are more directly involved in daily operations. This is a decentralization mechanism known as "management by objectives." Thompson maintains that subordinates are less willing to assume authority when they perceive decision errors as being too costly and when they feel that decisions are contingent upon too many factors that are beyond their control. To minimize the costs associated with poor decisions, subordinates form coalitions with other decision-makers within and/or outside the organization.

Stark concludes the chapter by noting that private sector organizations are moving more in the direction of decentralization while government organizations have become more centralized. Private organizations tend to be more efficient because their survival is tied more to performance. Consequently, indicators of success and failure are monitored routinely.

Key Learning Objectives

After a thorough reading of Chapter 20, you should be able to:

1. identify the basic characteristics of formal organizations.
2. describe how the creation of divisional systems enabled Moltke to maintain control over an expanding military organization.
3. note how the creation of geographical divisions made long rail lines more effective.
4. indicate how Swift utilized vertical integration and functional divisions to revolutionize his meat-packing business.
5. identify problems generated by the spoils system that are potentially corrected by the civil service system.
6. explain Weber's concept of rational bureaucracy.
7. distinguish rational and natural system approaches to the study of formal organizations.
8. provide examples of how organizational survival is enhanced through goal displacement.
9. indicate how goal conflict and informal networks impact organizational efficiency.
10. explain why highly diversified organizations are more efficiently controlled through a system of autonomous divisions.
11. state the major tenets associated with Blau's theory of administrative growth.
12. indicate why the "management by objectives" approach is an example of decentralization.
13. describe the interrelationship among discretion, responsibility and authority.
14. specify conditions that constrain a subordinate's willingness to exercise discretion.
15. explain why private organizations may be more vulnerable than government organizations.

Chapter Outline

I. The Rise of Formal Organizations
   A. Divisional Systems and the Military: Moltke
      1. Moltke created a highly trained, interchangeable staff.
      2. The army was divided into standardized units (the divisional system) that were managed by Moltke's General Staff.
   B. Geographical Divisions and Railroads: McCallum
      1. Short-distance lines were more efficient than long-distance lines.
      2. Long-distance lines were divided into a series of short-distance segments and placed under the authority of a supervisor who would report to the home office.
   C. Functional Divisions and the Meat-Packing Industry: Swift
      1. The key components of the meat-packing business were identified (functional divisions).
      2. Managers were placed in charge of each component so that activities could be controlled from start to finish (vertical integration).
   D. Civil Service and the Government
      1. Supporters of successful political candidates were rewarded with government appointments under the spoils system.
      2. The spoils system generated unstable administrations that were potentially staffed with incompetent persons.
      3. Under the civil service system, government workers are hired and promoted on the basis of technical qualifications and merit.

II. Weber and Rational Bureaucracies
   A. A System of Rational Rules and Procedures
   B. Bureaucracies and Specialization
   C. Hierarchical Chains of Command
   D. Monitoring Progress with Records and Files
   E. Goal-Oriented Organizations

III. Rational and Natural System Approaches to Formal Organizations
   A. Rational System Approach: Goal Attainment
   B. Natural System Approach: Informal Mechanisms of Control
      1. Stated goals may be replaced when an organization's survival is threatened (goal displacement).
      2. The members of an organization may not ascribe to the same goals (goal conflict).
      3. Persons may use informal networks to circumvent the chain of command (informal relations).

IV. Increasing Diversification and Decentralization
   A. Autonomous Divisions: The Du Pont Experience
      1. Functional divisions represent an effective approach to maintaining

control over formal organizations that pursue one major task.

2. Du Pont, a manufacturer of gunpowder and explosives, experienced tremendous expansion during WWI.
3. Following WWI Du Pont explored and developed other markets.
4. Since the needs of each industry were different, each product division was treated as a separate business, and a functional divisional system was implemented within each industry (autonomous divisions).

B. A Theory of Administrative Growth: Blau
1. Management costs increase as an organization increases in size and becomes involved in more diverse activities.
2. Smaller organizations are more cost effective and efficient.

C. Decentralization and Management by Objectives
1. As organizations grow, top management is more removed from direct supervision of the organization's activities. Consequently, management delegates decision-making authority to subordinates who are more directly involved in daily operations (management by objectives).
2. Subordinates are less willing to assume authority when decisions are perceived as too costly and when decisions are contingent on too many functions that are beyond their control.
3. Subordinates form coalitions with other decision-makers in order to minimize the costs associated with poor decisions.

D. Private Sector and Public Organizations
1. Private sector organizations are moving more in the direction of decentralization while government organizations have become more centralized.
2. Private organizations tend to be more successful because their survival is tied more to performance.

Key Terms

Based on your reading of Chapter 20, you should be able to define and illustrate the key sociological concepts listed below. Page numbers are provided in parentheses as reference points.

Formal Organization (570)
Rational Organization (570)
Divisional System (572)
Geographical Divisions (573)
Functional Divisions (573)
Vertical Integration (573)
Spoils System (576)
Civil Service (576)
Bureaucracy (576)
Rationality (576)
Management (577)
Rational System Approach (578)
Natural System Approach (578)
Goal Displacement (579)
Goal Conflict (579)
Span of Control (582)
Diversified Organization (583)
Autonomous Divisions (583)
Decentralization (583)
Blau's Administrative Theory (583)
Management By Objectives (584)
Discretion (584)

Bureaucrat (586)

Key Research Studies

Listed below are key research studies cited in Chapter 20. Familiarize yourself with the major findings of these studies. Page references are provided in parentheses.

Selznick: conducted study of the Tennessee Valley Authority. Goal displacement strategies were utilized by the agency to improve public image (579)

Blau: developed a theory linking organizational growth, differentiation and administrative costs. The theory argues that the cost of management increases as an organization increases in size and becomes involved in more diverse activities (583-584).

Drucker: writings have been associated with the "management by objectives" approach to management. This perspective maintains that some of the decision-making authority in large organizations should be delegated to persons more directly involved in the organization's daily operations (584).

Thompson: developed theoretical propositions concerning the use of discretion in decentralized organizations. Subordinates are less willing to assume authority when they perceive decision errors as being too costly and when they feel decisions are contingent upon too many factors that are beyond their control (584-585).

Info Trac Search Words

Enter these search terms to conduct more extensive investigations of key topics introduced in Chapter 20.

Patronage, Political
Subdivisions: Analysis

Bureaucracy
Select one of the articles listed. What characteristics of bureaucracy are addressed in your study? Does the author of your study take more of a rational or natural systems approach to the study of bureaucracy?

Management
Subdivisions: Models
Review the studies listed and select one that interests you. What management model is being presented? Does the model promote increasing decentralization or centralization of authority? How does the model guage efficiency?

Informal Networks

Management By Objectives

Multiple Choice

Answers and page references are provided at the chapter end.

1. By the __________ century, led by one person, military experts clearly recognized that armies had grown too large and could not be led effectively by one person.
   a. twenty-first
   b. twentieth
   c. nineteenth
   d. eighteenth
   e. seventeenth

2. Since the activities of formal organizations are based on a system of logical rules, formal organizations are also known as
   a. irrational organizations.
   b. rational organizations.
   c. autonomous divisions.
   d. dependent divisions.
   e. diversified organizations.

3. A basic characteristic of a formal organization is
   a. a clear statement of goals.
   b. trained leaders.
   c. clear lines of authority and communication.
   d. a system of records.
   e. all of the above

4. Formal organizations are relatively new phenomena as large organizations have existed for only about
   a. 100 years.
   b. 75 years.
   c. 50 years.
   d. 25 years.
   e. 10 years.

5. Stark reviews the growth of formal organizations in these areas
   a. business, religion and education.
   b. government, medicine and business.
   c. religion, the military and the family.
   d. the military, business and government.
   e. education, medicine and government.

6. The development of divisional systems is associated with
   a. McCallum.
   b. Moltke.
   c. Swift.
   d. Napoleon.
   e. Weber.

7. Moltke discovered that he could maintain control of the army with
   a. a highly trained staff.
   b. a staff that could be interchangeable.
   c. better armed, braver soldiers.
   d. a and b
   e. a and c

8. Wellington and Moltke created small, identically structured fighting units as a way to gain more control of their armies. This is known as
   a. vertical integration.
   b. horizontal integration.
   c. divisional systems.
   d. functional divisions.
   e. geographical divisions.

9. Moltke standardized smaller fighting units by making them similar in terms of
   a. size.
   b. makeup.
   c. training.
   d. structure.
   e. all of the above

10. The problem of inefficient long-distance rail lines was resolved by
   a. Swift.
   b. McCallum.
   c. Weber.
   d. Wellington.
   e. Moltke.

11. McCallum decided to break long-distance rail lines into a series of short-distance lines, each headed by a supervisor. McCallum is credited with having devised__________ divisions.
   a. geographical
   b. functional
   c. independent
   d. dependent
   e. autonomous

12. In this company the local manager reports to the district manager who in turn reports to the main corporate office. This is an example of a/an
    a. spoils system.
    b. civil service system.
    c. organizational chart.
    d. goal displacement strategy.
    e. decentralized company.

13. Swift created a purchasing, packing, shipping, and marketing department for his meat-packing company. The company was therefore subdivided on the basis of task. This is an example of
    a. horizontal divisions.
    b. rational divisions.
    c. formal divisions.
    d. functional divisions.
    e. geographical divisions.

14. Identifying and controlling each step of a process from start to finish is known as __________ integration.
    a. positive
    b. negative
    c. vertical
    d. horizontal
    e. centralized

15. Political leaders may reward their key supporters by giving them important positions in government. This is known as
    a. goal dispowerment.
    b. political decentralization.
    c. political centralization.
    d. the spoils system.
    e. the civil service system.

16. Recruitment on the basis of educational and occupational qualifications is a key feature associated with
    a. the spoils system.
    b. civil service.
    c. management by objectives.
    d. nepotism.
    e. discretion

17. To work for the U.S. Postal Service a person must take the postal exam and receive a certain score. This is an example of
   a. vetical integration.
   b. the spoils system.
   c. the “old boy” network.
   d. civil service.
   e. span of control.

18. Each of the characteristics listed below is a characteristic of Weber’s Rational Bureaucracy except
   a. promotion is based on favoritism.
   b. a bureaucracy includes areas of specialization.
   c. the chain of command is hierarchical.
   d. decision-making is based on rules and procedures.
   e. leaders must be selected and adequately trained.

19. Weber argued that bureaucratic organizations
   a. represent an attempt to regulate organizational behavior through the use of rules and reason.
   b. were a rational solution to the problems presented by increasing organizational size.
   c. were a rational product of social engineering.
   d. were technically superior to any other organizational form.
   e. all of the above

20. A natural systems approach to formal organizations addresses
   a. goal displacement.
   b. goal conflict.
   c. informal networks.
   d. each of the above
   e. none of the above

21. In response to concern over its scope of influence, the Tennessee Valley Authority (TVA) later announced that its main focus would be rural electrification and resource management. This is an example of
   a. goal decentralization.
   b. goal management.
   c. goal displacement.
   d. goal overload.
   e. goal strain.

22. The DuPont experience indicates that control of large corporations involved in diverse markets is best maintained through the use of
    a. geographical divisions.
    b. autonomous divisions.
    c. functional divisions.
    d. natural divisions.
    e. civil divisions.

23. Research suggests that an effective leader should have no more than seven subordinates reporting directly to him or her. This is an illustration of
    a. the span of control.
    b. sphere of influence.
    c. natural management.
    d. management by control.
    e. management by objectives.

24. __________ divisions work well in organizing the activities of companies with a limited market, but diversified organizations rely more on __________ divisions to effectively coordinate the organization's activities.
    a. Functional; geographic
    b. Geographic; functional
    c. Autonomous; geographic
    d. Autonomous; functional
    e. Functional; autonomous

25. Autonomous divisions and the "management by objectives" approach are examples of
    a. centralization.
    b. decentralization.
    c. anomie.
    d. organizational spoils.
    e. bureaucratic inefficiency.

26. As organizations grow a greater proportion of the organization's resources must be directed to managerial costs. This is a key concept associated with
    a. Drucker's "management by objectives" perspective.
    b. Blau's theory of administrative growth.
    c. the theory of autonomous divisions.
    d. Thompson's theory of discretion.
    e. the theory of natural systems.

27. According to Blau's theory of administrative growth
    a. organizational growth stimulates differentiation.
    b. differentiation stimulates administrative growth.
    c. differentiation restricts and limits administrative growth.
    d. a and b
    e. a and c

28. In a decentralized organization, __________ is defined as the freedom to make choices and decisions.
  a. responsibility
  b. authority
  c. autonomy
  d. rationality
  e. discretion

29. The transfer of some key decision-making responsibilities from top level management to subordinates who are familiar with the organization's daily operations is a key feature of this approach to management.
  a. management by spoils
  b. civil management
  c. management by objectives
  d. displacement management
  e. scientific management

30. Thompson argues that subordinates are less likely to exercise discretion when
  a. they perceive decision errors to be too costly.
  b. when they feel that decisions are contingent upon too many factors that are beyond their control.
  c. when the decision must be made quickly, within a short time frame.
  d. a and b.
  e. b and c.

True/False

Answers and page references are provided at the chapter end.

1. Since formal organizations are based on logical rules, they are often described as rational organizations.

2. Moltke created geographical divisions to gain control over his large military organization.

3. The principle of vertical integration is being applied when each step of a company's operations is being controlled and monitored.

4. The spoils system is to civil service as merit is to reward.

5. The rational bureaucracy concept is most closely associated with the thought of Durkheim.

6. Within a rational bureaucracy, authority and power are portrayed in terms of a hierarchical chain of command.

7. Natural system approaches to formal organizations place more emphasis on an organization's official, intended characteristics.

8. The DuPont Corporation resolved its managerial control problems by creating autonomous divisions within the corporation.

9. Blau maintains that managerial costs decrease relative to other costs, as organizations grow larger.

10. Within a decision-making context, discretion hinges on responsibility and authority.

Short Answer Questions

These short answer questions are provided to test your knowledge of the basic sociological concepts presented in Chapter 20. Page references for answers are included at the chapter end.

1. Identify five basic characteristics of formal organizations.

2. How did Moltke solve the problems of commanding a large army?

3. McCallum created a geographical division system to improve the efficiency of long-distance rail shipping and travel. How did this system operate?

4. Explain why vertical integration and functional divisions are complimentary mechanisms enhancing organizational control.

5. When Weber argued that bureaucracies are rational systems, what did he mean?

6. Distinguish the rational system and natural system approach to the study of formal organizations.

7. Provide examples of goal displacement and goal conflict within formal organizations.

8. Why was top management better able to gain more efficient control over the activities of the Du Pont Corporation when they decided to restructure the corporation into autonomous divisions?

9. Identify two of the major propositions associated with Blau's administrative theory.

10. How is "the management by objectives" approach to management an example of decentralization.

Essay Questions

These questions are designed to test you understanding of key sociological concepts presented in Chapter 20 and you ability to apply these insights to concrete situations.

1. Design your own business organizational chart utilizing the principles of vertical integration and function divisions.

2. A college graduate is trying to land that "first job." Which is more important, merit or the spoils system. Justify your answer.

3. Explain why a college or university is an example of a rational bureaucracy.

4. Why are rational and natural system approaches to the study of formal organizations considered complimentary rather than conflicting approaches?

5. What factors have contributed to the increasing decentralization of private sector organizations and the increasing centralization of government organizations?

Answers

Multiple Choice

1. c (569)
2. b (570)
3. e (570)
4. a (570)
5. d (571)
6. b (571-572)
7. d (571)
8. c (572-573)
9. e (572)
10. b (573)
11. a (573)
12. c (573)
13. d (574-575)
14. c (574)
15. d (576)
16. b (576)
17. d (576)
18. a (576)
19. e (576-577)
20. d (578-580)
21. c (579)
22. b (581-583)
23. a (582)
24. e (581, 583)
25. b (583-584)
26. b (583)
27. d (593)
28. e (584)
29. c (584)
30. d (584-585)

True/False

1. T (570)
2. F (571-573)
3. T (574-575)
4. F (576)
5. F (576)
6. T (576)
7. F (578)
8. T (583)
9. F (583)
10. T (584)

Short Answer

1. (570)
2. (571-573)
3. (573)
4. (574-575)
5. (576-578)
6. (578)
7. (578-579)
8. (582-583)
9. (583)
10. (584-585)

CHAPTER 21

SOCIAL CHANGE AND SOCIAL MOVEMENTS

Extended Abstract

In this chapter Stark combines aspects of the collective behavior and the resource mobilization approachs to social movements to provide an analysis of key events associated with the Civil Rights Movement. Two key events are the 1955-1956 Montgomery, Alabama bus boycott and the 1964 Mississippi "Freedom Summer."

Social movements are examples of collective, organized behavior that is designed to bring about change or oppose change. The collective behavior approach to social movements portrays social movements as responses to deeply held grievances. Social movements have a strong ideological and emotional base. The resource mobilization approach is a more rational approach. Rational planning, effective leadership, financial resources and a high degree of personal commitment are key ingredients necessary to insure a social movement's success.

In order to provide a more comprehensive account of how social movements arise and whether they succeed or fail, Stark develops eight propositions based on key features of the collective movement and resource mobilization approaches. Four of the propositions address why social movements occur, and four offer an explanation for a movement's success or failure. The four propositions explaining the emergence of social movements address such issues as a shared grievance, reasonable hope of success, a catalytic event and the presence of strong social networks. The four propositions associated with success focus on the mobilization of personal and financial resources, the ability to combat opposition movements, help from powerful, external allies, and finally the movement's concerns are embodied in multiple organizations.

Having provided this theoretical framework, Stark demonstrates how the social movement propositions provide a better understanding of the origin of the Montgomery bus boycott, its success and the continued success of the Civil Rights Movement. The shared grievance for African Americans was the long history of racial discrimination and inequality. African Americans experienced economic, political and personal domination in "white" America. African American men were employed in low paying jobs, and many African American women were employed as domestics. Few were registered to vote, and life in segregated America was separate but not equal. By the mid-1950s there was reason to believe that things would be better. The 1954 Supreme Court ruling on school segregation represented hope that life would be better. Social scientists argue that social crises arise when the gap between the rate of actual change and deserved change widens. This is known as the J-curve theory of social crisis. Although the Supreme Court ruling represented hope, the inequalities and discrimination of segregated America were becoming increasingly more intolerable. The precipitating event came in 1955 when Rosa Parks violated Alabama's bus segregation law and was arrested. This event unified the African American community. It was now evident that change was necessary. As a result of strong network ties linking African American professors, clergy and members of the NAACP, the African American community responded quickly and organized an effective boycott of the Montgomery public transportation system.

Personal and financial resources were mobilized quickly as Martin Luther King, Jr., and others assumed important leadership positions. Churches served as strategic mass meeting venues, and members of the African American community donated their time and financial resources to support the boycott. Soon the Montgomery Improvement Association (MIA) was formed, and Martin Luther King, Jr. was elected president. At that point the organization decided to maintain an ongoing non-violent protest of the bus segregation law. The movement was met by a significant white countermovement. White Citizens' Councils were formed, and African American harassment intensified. However, the MIA began to receive support from sources external to the community. Church groups outside the South sent money, and the boycott received national press coverage thus building additional outside support for the boycott. The issues associated with the boycott received national attention, and the U.S. Supreme Court declared the bus segregation law unconstitutional in November 1956, almost a year after Rosa Park's protest and arrest. However, the victory was only a partial victory. Segregation policies still structured many areas of daily life such as restaurants, hotels and hospital care. African American leaders met in Atlanta to form the Southern Christian Leadership Conference (SCLC) and address more broad-based civil rights violations. Other Civil Rights organizations, like the Student Non-Violent Coordinating Committee (SNCC) were formed, and the Civil Rights Movement continued on a national scale

In the last part of the chapter, Stark provides a closer look at the SNCC's 1964 "Freedom Summer." The event was organized as a Mississippi voter registration campaign. Organizers believed that if more African Americans were registered to vote, racial inequality in Mississippi could be reduced. Students from more affluent backgrounds attending prestigious schools were recruited to participate in the project. It was assumed that the families of these students' would have powerful political and economic connections and that student mistreatment would not be tolerated. The workers encountered substantial violence, but their efforts and mistreatment received national media attention. By the next summer (1965) the federal Voting Rights Act was signed.

A follow-up study of the "Freedom Summer" volunteers was conducted in the mid-1980s by McAdam, who believed that many of the key figures associated with contemporary social movements had also participated in the 1964 "Freedom Summer" campaign. He was able to locate the applications for 959 students who had participated in the campaign and for another 300 who volunteered, but did not participate. These data were analyzed, and follow-up interviews with forty past volunteers and forty non-participants were conducted. McAdam discovered that those who participated were free to participate. They were biographically available. They did not have other ties or obligations that would prevent their participation. Non-participants included more female students, younger students and students who had encountered parental opposition. Those who participated had developed strong ties to Civil Rights organizations prior to volunteering and believed that they could make a difference. The majority of the volunteers also had a friend who had volunteered. Thus, persons participate in social movements as a result of the development of strong social ties and networks. The 1984 follow-up interviews revealed that many of the volunteers continued to take a more radical position on political and social issues and were involved in various social movements. They also were more likely to be divorced and to experience a degree of social isolation.

Key Learning Objectives

After a thorough reading of Chapter 21, you should be able to:

1. provide a brief outline of key events associated with the beginning of the Civil Rights Movement.
2. identify major tenets associated with the collective behavior approach to social movements.
3. explain the resource mobilization approach to social movements.
4. state and evaluate Stark's eight propositions relating to social movement occurrence and success.
5. provide examples of African American experiences of economic, political and personal domination prior to the Civil Rights Movement.
6. explain the J-Curve theory of social crisis.
7. identify crucial events that precipitated the Montgomery bus boycott.
8. distinguish important network ties that enabled the African American community to mobilize resources and organize an effective bus boycott.
9. identify critical internal and external factors associated with the Montgomery bus boycott.
10. document the proliferation of civil rights organizations following the Montgomery bus boycott.
11. identify the purpose of the Student Non-Violent Coordinating Committee's 1964 "Freedom Summer."
12. indicate what was distinctive about the participants in the 1964 "Freedom Summer" voter registration campaign.
13. state the key findings of McAdam's study of the "Freedom Summer" volunteers.
14. define biographical availability and provide an example.
15. explain how persons can be part of their social environment and history and shape their environment and the course of history.

Chapter Outline

I. Analyzing Social Movements
   A. Rosa Parks Challenges Segregation
   B. Organizations Designed to Cause or Prevent Change (Social Movements)
   C. Collective Behavior Approach to Social Movements
      1. Social movements are group responses based on deeply felt grievances.
      2. A social movement's goals are ideologically based.
   D. Resource Mobilization Approach to Social Movements
      1. Social movements must have adequate human and financial resources.
      2. Rational planning and effective leadership are critical to a social movement's success.
   E. Stark's Social Movement Model Propositions
      1. Persons share grievances that they desire to address.
      2. Persons are hopeful that things will be better.
      3. A key event stimulates or precipitates a collective response.
      4. Persons are drawn to a movement through strong social network ties.
      5. Human and financial resources must be mobilized if a movement is to be successful.
      6. Successful movements are able to withstand countermovements.
      7. Successful movements are able to mobilize external support.
      8. Movements experiencing broad-base support generate additional organizations designed to address the shared grievance.

II. The Montgomery Bus Boycott and Stark's Social Movement Propositions
   A. Shared Grievances: Economic, Political and Personal Domination
      1. African American men and women were often employed in low status jobs.
      2. Voter registration was low among African Americans.
      3. Life in segregated America was separate but not equal.
   B. Indications of Hope
      1. The U.S. Supreme Court (1954) rules that school segregation is unconstitutional.
      2. A social crisis tends to arise when the gap between the rate of actual and desired change widens (J-curve theory of social crisis).
   C. The Precipitating Event: The Arrest of Rosa Parks
      1. Rosa Parks challenges the Alabama bus segregation law and is arrested (December 1955).
      2. This event angers the African American community and signals the need for a change.
   D. Social Networks and the Montgomery Bus Boycott
      1. Key figures in the Montgomery African American community mobilize resources to organize a boycott.
      2. Strong social networks based on friendship ties, church ties and organizational ties (NAACP) exist.
   E. Mobilizing Resources: The Montgomery Improvement Association (MIA).

1. Churches function as strategic mass meeting venues.
2. Members of the African American community donate time and financial resources to support the boycott.
3. The Montgomery Improvement Association (MIA) is formed to maintain the boycott, and Martin Luther King, Jr. is elected organization president.
4. MIA dedicates itself to a program of non-violent protest.

F. The Countermovement: The White Citizens' Councils
1. White Citizens' Councils form as a white opposition movement.
2. Harassment of African Americans intensifies and becomes more violent.

G. The Rise of External Support
1. Church groups outside Montgomery and outside the South began to provide financial support.
2. The boycott receives national media coverage.
3. The U.S. Supreme Court declares the Alabama bus segregation law unconstitutional (November 1956).

H. Incomplete Mission: The Proliferation of Civil Rights Organizations
1. U.S. Supreme Court decision only addresses segregated bussing.
2. Segregation policies still structure many areas of daily life such as restaurants, hotels and hospital care.
3. African American leaders meet in Atlanta and form the Southern Christian Leadership Conference (SCLC) to address broader Civil Rights violations.
4. Students are involved in the Civil Rights Movement through the Student Non-Violent Coordinating Committee (SNCC).

III. The 1964 Mississippi "Freedom Summer"
A. Mission and Goal: Voter Registration
B. Participants: Student's From More Affluent Backgrounds
1. It is assumed that students from more affluent backgrounds have strong network links to powerful political and economic resources.
2. It is also assumed that the mistreatment of students with important network ties will not be tolerated.

C. The Outcome
1. Students encountered substantial opposition and violence.
2. The students' efforts and mistreatment receives national attention.
3. The federal Voting Rights Act is signed the following summer (August 1965).

IV. Studying the "Freedom Summer" Volunteers (McAdam)
A. "Freedom Summer" Volunteers and Their Current Social Movement Involvement
B. "Freedom Summer" Volunteer Applications and Follow-up Interviews
C. Important Study Findings

1. Students who participated in "Freedom Summer" did not have other ties or obligations preventing their participation. They were biographically available.
2. Non-participants included more female students, younger students and students who encountered more parental opposition.
3. Participants developed strong ties to Civil Rights organizations prior to volunteering.
4. Participants believed they could bring about change.
5. Most participants had a friend who also participated.

6. The 1984 follow-up interviews indicated that many of the volunteers have remained politically active, take more radical stances and are involved in contemporary social movements.
7. The follow-up interviews also indicate that the volunteers were more apt to be divorced and to experience social isolation.

Key Terms

Based on your reading of Chapter 21, you should be able to define and illustrate the key sociological concepts listed below. Page numbers are provided in parentheses as reference points.

Social Movement (594)
Collective Behavior Approach (595)
Resource Mobilization Approach (595)
Countermovements (596)
Revolution of Rising Expectations (599)
J-Curve Theory of Social Crisis (599)
Precipitating Event (599)
Biographical Availability (608)

Key Research Studies

Listed below is the key research study cited in Chapter 21. Familiarize yourself with the major findings of this study. Page references are provided in parentheses.

McAdam: conducted a study of "Freedom Summer" volunteers. Program applications were reviewed in order to specify identifying characteristics of participants and non-participants. Participants were biographically available, believed they could bring about change and had developed strong social ties with Civil Rights organizations and with other persons volunteering. The follow-up survey revealed that the "Freedom Summer" volunteers remained politically active, took more radical stances and were involved in contemporary social movements. Participants were also more likely to be divorced and to have experienced a degree of social isolation (607-611).

Info Trac Search Words

Enter these search terms to conduct more extensive investigations of key topics introduced in Chapter 21.

King, Martin Luther, Jr.
Subdivisions: Social Policy

Collective Behavior
Subdivisions: Analysis
Review the studies cited and select one that interests you. What type of collective behavior is described in the study you selected? Is the behavior ideologically based? Can a shared grievance be identified? How might the issue you selected be addressed from a resource mobilization perspective?

Civil Rights Movements
Subdivisions: History
Select one of the studies cited. What aspect of the Civil Rights Movement is addressed? Can Stark's propositions be applied to this aspect of the Civil Rights Movement? If so, what insights are gained by applying this theoretical framework?

Churches and Civil Rights

Freedom Summer

Multiple Choice

Answers and page references are provided at the chapter end.

1. Rosa Parks was arrested for violating the Montgomery
   a. restaurant segregation law.
   b. bus segregation law.
   c. hotel segregation law.
   d. movie theatre segregation law.
   e. hospital segregation law.

2. The Montgomery Improvement Association was an example of a/an
   a. housing authority.
   b. revolution.
   c. social movement.
   d. ideology.
   e. job training program.

3. The collective behavior approach to social movements addresses
   a. ideology and shared grievances.
   b. rational planning.
   c. effective leadership.
   d. financial resources.
   e. all of the above

4. Rational planning and effective leadership are primary emphases of the __________ approach to social movements.
   a. collective behavior
   b. natural system
   c. bureaucratic
   d. rational choice
   e. resource mobilization

5. For a social movement to occur
   a. people must share a grievance.
   b. some hope for success must be present.
   c. a precipitating event may serve as a catalyst for change.
   d. people are recruited to the movement through social networks.
   e. all of the above

6. Effective leadership, the attraction of committed members and the ability to secure resources are primarily associated with this feature of social movements.
   a. hope
   b. mobilization of people and resources
   c. shared grievance
   d. countermovement
   e. precipitating event

7. The U.S. Supreme Court ruling declaring school segregation unconstitutional best illustrates this aspect of social movements.
   a. an internal force
   b. countermovement
   c. hope
   d. precipitating event
   e. shared grievance

8. African American women in the 1940s and early 1950s were most likely to be employed as
   a. school teachers.
   b. nurses.
   c. seamstresses.
   d, domestics.
   e. librarians

9. The economic, political and personal domination experienced by African Americans prior to the Civil Rights Movement are cites by Stark as examples of
   a. external allies.
   b. countermovements.
   c. shared grievances.
   d. networks of attachments.
   e. hope.

10. Low African American voter registration in many Southern counties in the early 1950s is an example of
   a. geographic domination.
   b. economic domination.
   c. personal domination.
   d. political domination.
   e. educational domination.

11. At the time of the Civil Rights Movement, African Americans were expected to greet whites in a formal manner; whereas, whites were more likely to greet African Americans by their first name. This is an example of
   a. personal domination.
   b. religious domination.
   c. economic domination.
   d. educational domination.
   e. revenue domination.

12. The "revolution of rising expectations" is most likely to be associated with this social movement proposition.
   a. shared grievance
   b. hope
   c. countermovement
   d. mobilization of people and resources
   e. recruitment through social networks

13. According to the J-curve theory of social crisis, anger and dissatisfaction __________ as the gap between the hope for change and the actual rate of change __________.
   a. decrease, increases
   b. increase, decreases
   c. increase, stays the same
   d. increase, increases
   e. decrease, stays the same

14. The belief that social change will be rapid is known as
   a. great expectations.
   b. the idea of progress.
   c. a revolution of rising expectations.
   d. collective mobilization.
   e. collective advance.

15. The catalyst or dramatic action which prompts the development of a social movement is known as a

a. primary event.
b. secondary event.
c. rational event.
d. natural event.
e. precipitating event.

16. Typically social movements

a. are founded by a single person.
b. are created by a group.
c. grow through social attachments.
d. a and b
e. b and c

17. With regard to the early development of the Montgomery bus boycott, friends, church and committee memberships are examples of

a. support from external allies.
b. countermovements
c. network ties.
d. biographical availability.
e. precipitating events.

18. For the Montgomery bus boycott to be successful, strong leadership and a large group of persons committed to the cause would be necessary. This illustrates the following social movement proposition.

a. precipitating event.
b. mobilizing of people and resources.
c. proliferation of organizations.
d. hope.
e. support from external allies.

19. The Montgomery bus boycott was an effective protest tactic because

a. the majority of the bus drivers were African American.
b. the majority of the bus passengers were African American.
c. gas prices were high.
d. the city could not continue to maintain separate buses for African Americans and whites.
e. most people preferred to ride the bus rather than drive their car to work.

20. A major precipitating event for the development of white countermovements during the Civil Rights struggle was

a. the U.S. Supreme Court decision on school segregations.
b. the Voting Rights Act.
c. the Montgomery bus boycott.
d. Freedom Summer.
e. the integration of the military and professional sports.

21. The Montgomery Improvement Association received support from whites, churches in the North and South and the media. This illustrates the following social movement proposition.
    a. external opposition.
    b. external support/allies.
    c. hope.
    d. share grievance.
    e. precipitating events.

22. The creation of the Southern Leadership Conference (SCLC) and the Student Non-Violent Coordinating Committee (SNCC) best illustrate this social movement proposition.
    a. the proliferation of separate organizations
    b. biographical availability
    c. revolution of rising expectations
    d. precipitating events
    e. external oppositions

23. In Mississippi in the early 1960s
    a. there were no registered African American voters in five counties where African Americans were in the majority.
    b. half of African American homes lacked running water.
    c. two-thirds of African American homes lacked toilets.
    d. only 7 percent of African Americans had graduated form high school.
    e. all of the above.

24. In the early 1960s __________ percent of the African American adults in Mississippi were registered to vote.
    a. less than five
    b. approximately seven
    c. twenty
    d. thirty-three
    e. fifty

25. "Freedom Summer" participants included
    a. a sizable number of white students.
    b. students from affluent families.
    c. students from prestigious colleges and universities.
    d. a and b.
    e. a, b and c.

26. Within two months after the "Freedom Summer" campaign began
    a. African Americans received the right to vote.
    b. bus segregation in Mississippi was declared unconstitutional.
    c. the Mississippi state legislature passed the Mississippi Civil Rights Act.
    d. four Civil Rights workers had been killed and thirty-seven churches had been bombed.
    e. all of the above

27. Many of the students who participated in the "Freedom Summer" campaign were able to do so because they did not have family or job commitments and responsibilities that limited their participation. They were therefore
    a. biographically available
    b. geographically available.
    c. ideologically available.
    d. politically available.
    e. c and d.

28. Many of the "Freedom Summer" volunteers were the children of parents who were also active in more liberal political causes. Researchers have labeled this phenomenon the
    a. white diaper phenomenon.
    b. black diaper phenomenon.
    c. red diaper phenomenon.
    d. activist family syndrome.
    e. radical family syndrome.

29. Which of the following findings from McAdam's study of the "Freedom Summer" participants is false.
    a. Those who applied but failed to participate were primarily older, male students who encountered parental opposition.
    b. The participants believed in racial equality and felt that their participation would help bring about racial equality.
    c. Many of the participants had ties to Civil Rights organizations and had been involved in other forms of activism.
    d. Two-thirds of the volunteers had a friend that was also a volunteer.
    e. Each statement is true.

30. McAdam's follow-up interviews revealed that presently those that had participated in "Freedom Summer"
    a. were no longer politically active.
    b. were still politically involved and were active in different social movements.
    c. had experienced high divorce rates and a significant degree of social isolation.
    d. a and c
    e. b and c

True/False

Answers and page references are provided at the chapter end.

1. Social movements can arise to cause or prevent social change.

2. The resource mobilization approach to social movements places the heaviest emphasis on leadership.

3. People are primarily recruited to social movements on the basis of the movement's ideology.

4. Prior to the Civil Right's Movement, African American women were less likely to be employed than were white women.

5. The "sit in" at the Woolworth's lunch counter in Greensboro, NC was the event that precipitated the Civil Rights Movement.

6. The church is an example of a social network tie within the African American community that played a key role in the Civil Rights Movement.

7. The Montgomery bus boycott was effective because the majority of passengers were white.

8. The White Citizens' Council was an example of a Civil Rights countermovement.

9. The major focus of the Mississippi "Freedom Summer" campaign was voter registration.

10. McAdam discovered that most of the students who volunteered for the "Freedom Summer" campaign had little or no contact with the Civil Rights Movement prior to volunteering.

Short Answer

These short answer questions are provided to test your knowledge and understanding of the basic sociological concepts presented in chapter 21. Page references for answers are included at the chapter end.

1. What event appears to have precipitated the Civil Rights Movement?

2. Compare and contrast the collective behavior and resource mobilization approaches to social movements.

3. What four conditions must be present for a social movement to occur according to Stark?

4. According to Stark, what four features contribute to a social movement's success?

5. Just prior to the beginning of the Civil Rights Movement, how were African American being dominated economically, politically and personally?

6. Explain the J-curve theory of social crisis.

7. Describe how strong network ties played a key role in the organization of the Montgomery bus boycott?

8. Are countermovements social movements?

9. What social characteristics distinguished those who actually participated in the "Freedom Summer" from those who applied but failed to participate?

10. Identify some of the key findings of McAdam's follow-up interviews with "Freedom Summer" participants.

Essay Questions

These questions are designed to test your understanding of key sociological concepts presented Chapter 21 and your ability to apply these insights to concrete situations.

1. Provide a critical evaluation of the Civil Rights Movement. What aspects of racial inequality have been resolved, and what issues remain?

2. How would functionalist and conflict theorists differ in their interpretation of social movements?

3. Utilize Stark's social movement propositions to evaluate the rise and success of the environmental movement in the U.S.

4. Provide an example of a contemporary social movement and its corresponding countermovement.

5. How would biographical availability vary by race, class, age and gender?

Answers

Multiple Choice

1. b (593-594)
2. c (594, 601-602)
3. a (595)
4. e (595)
5. e (596)
6. b (596)
7. c (596, 598-599)
8. d (597)
9. c (596-598)
10. d (597)
11. a (597)
12. b (599)
13. d (599)
14. c (599)
15. e (599)
16. e (600)
17. c (600)
18. b (601)
19. b (602)
20. a (599; 602)
21. b (603-605)
22. a (605)
23. e (605)
24. b (605)
25. e (606)
26. d (606)
27. a (608)
28. c (608)
29. a (608-610)
30. e (610-611)

True/False

1. T (594)
2. T (595)
3. F (596)
4. F (597)
5. F (598-599)
6. T (600-601)
7. F (602)
8. T (602-603)
9. T (606)
10. F (609)

Short Answer

1. (593-594, 599-600)
2. (595-596)
3. (596)
4. (596)
5. (597 –598)
6. (599)
7. (600-602)
8. (596, 602-603)
9. (606, 608-610)
10. (610-611)